My Ranch, Too is an honest, plain-speaking book about a way of life and work that the author rightly understands as a vocation. From start to finish, Mary Flitner tells what she authentically knows, drawing from her own experience and the involvement of several generations of her family. She is always conscientiously practical, giving a full account of the rancher's responsibilities, from getting the calves safely born to getting the ranch safely passed to the next generation. She writes with the same care of the hardships and beauties of the work, of the pleasures and exasperations and sometimes the funniness of working with other people, and of the great love that makes sense of it all. This love is the book's subject and informing principle. The ability to express it—to show it, so to speak, in action—in the entire fabric of a book is not an ability a writer can acquire. She either has it or she doesn't. Mary Flitner has it.

Wendell Berry

My Ranch, Too

My Ranch, Too

A Wyoming Memoir

Mary Budd Flitner

Foreword by Teresa Jordan

UNIVERSITY OF OKLAHOMA PRESS : NORMAN

Publication of this book is made possible through the generosity of Edith Kinney Gaylord.

Library of Congress Cataloging-in-Publication Data

Name: Flitner, Mary Budd, 1942– author.
Title: My ranch, too : a Wyoming memoir / Mary Budd Flitner ; foreword by Teresa Jordan.
Description: Norman, OK : University of Oklahoma Press [2018]
Identifiers: LCCN 2018003408 | ISBN 978-0-8061-6058-0 (hardcover : alk. paper)
Subjects: LCSH: Ranchers—Wyoming—Biography. | Ranching—Wyoming.
Classification: LCC SF194 .F55 2018 | DDC 636/.0109787—dc23
LC record available at https://lccn.loc.gov/2018003408

The paper in this book meets the guidelines for permanence and durability of the Committee on Production Guidelines for Book Longevity of the Council on Library Resources, Inc. ∞

1 2 3 4 5 6 7 8 9 10

Contents

Illustrations

Photographs

Maps

◇— Foreword

Teresa Jordan

Loss is a dominant theme in the literature of the American West. "She's gone, goddam it! Gone!" says Uncle Jeb, a mountain man in A. B. Guthrie's *The Big Sky.* "The whole shitaree. Gone, by God, and naught to care savin' some of us who seen 'er new." The theme is particularly pronounced in the literature of cattle culture, where it has acquired a poignancy that can border on cliché. By example, in 1978, after *New Yorker* writer Jane Kramer published *The Last Cowboy,* the study of the ranch hand she rechristened Henry Blanton, one reviewer opined that "the mythical 'last cowboy' has . . . ridden off into the sunset so often that he's in danger of retina damage." The playwright and novelist Larry L. King knew whereof that reviewer spoke. Author of *The Best Little Whorehouse in Texas,* he admitted his own culpability. "I personally have discovered and recorded [the last cowboy] a half-dozen times and Larry McMurtry has made a career of it—to say nothing of William Eastlake, Max Crawford, Edwin Shrake and J. Frank Dobie."

The literature of ranching is not the same as that of the cowboy; the cowboy is a free spirit, in myth if not in reality, just passing through. Ranchers, for the most part, hope to stay. But the tone of ranching literature is often forlorn, and for good reason. Since 1935, more than 141 million acres of farm and ranch land have gone out of production. We are used to comparing land areas to Delaware or Rhode Island, but this loss is on another scale altogether, an acreage nearly as large as the state of Alaska. As farms and ranches have disappeared, so have farmers and ranchers, the population dropping by 90 percent over the same period, from nearly 32 million to under 3.2 million today.

Mary Flitner's memoir, *My Ranch, Too*, is set among these losses, but it stands apart from most ranch narratives because it is not an elegy so much as a hymn to resilience. Mary's great-grandfather, Daniel Bockins Budd, drove a herd of cattle into Wyoming Territory in 1878 and built a life there, and his initial holdings remain in the family's hands today. Mary's husband, Stan Flitner, comes from similar stock. His grandfather, Arthur Flitner, bought a ranch in northwest Wyoming in 1906: more than 110 years later, those original acres remain in the Flitner clan. Mary and Stan are now in their seventies; their children and grandchildren carry on the ranching tradition.

The ability of the Budd and Flitner ranches to bridge the nineteenth, twentieth, and twenty-first centuries is nothing short of remarkable. When nine out of ten families have left the land in the span of a single lifetime, the story of those who have survived generation after generation and still look to the future with a degree of confidence is something not only to celebrate but to study. What combination of grit, adaptation, creativity, and interpersonal dynamics has made such a thing possible? What lessons does such a journey have for the rest of us, on the land or off it, as we struggle to stay relevant in the face of ever-accelerating change? This is the trail that Mary charts in her riveting memoir.

———

The story begins in 1878 when Mary's great-grandfather learns that his thirty-five-year-old brother has died suddenly in northern California, near the border of Nevada. Daniel Bockins Budd left his job at a prison in Kansas to settle his brother's affairs and was surprised to learn that his sibling had become a successful cattleman out west. Daniel fell heir to almost eight hundred head of stock and set upon trailing them toward Omaha, where he felt they would bring a better price. He kept a journal of his experiences, diligently recording his expenses, his route, the obstacles he met along the way. We know that he spent seventy-five cents on dinner and gave five dollars to the poor. We know how long he rode along the Humboldt River; when he crossed the Mary's River; his tedious weeks heading north along the Green. We know about the cold muddy night he spent trying to pull sixty head out of a bog, and that he lost ten of them. We know that the Mormons at Fort Bridger denied him winter ground for his herd, and that he pushed on with increasing urgency as winter descended. We sense his relief, on November 13, 1879, when he was finally able to write, "On Big Piney . . . up Green River . . . Winter quarters." We learn that he boarded a train back to Kansas to rejoin his family, doubting that the cattle would survive the winter, and that when he returned in the spring to find them fat and shiny, he was so relieved he decided to settle not only the cattle but his family as well in Wyoming Territory.

But the journal leaves out as much as it reveals. We don't know why Daniel lost touch with his brother. We don't know how a prison guard transformed himself into a trail boss; or how Daniel cajoled his men into staying with him after they threatened to quit; or what words passed between him and the Mormons at Fort Bridger; or how he talked his wife into leaving the comforts of Kansas for the unknown privations of Wyoming Territory. As Mary notes, no matter how great the obstacles her great-grandfather encountered, he "does not express anger, disappointment or fear in his words, only a factual accounting of what could have been a near defeat. . . . After all, the journal wasn't a story book; it was a business account."

In this, the journal is not unlike many accounts of family ranching, most especially the rare stories of family success. We know the obstacles, and we can see that the family overcame them. But we seldom know *how,* or what particulars of human behavior determined their longevity. Daniel Budd recorded all the details we need to chart his journey on a map, but his account, like a map itself, is two-dimensional. Perhaps it was Mary's frustration with everything left unsaid that motivated her to dig deeper into the succeeding generations, to plumb that mysterious third coordinate, the human dimension, that gave her family staying power when so many others had to leave.

———

Some of the ground traversed here is familiar from other ranch accounts, above all the essence of the work itself: the art of reading cattle, of choreographing a successful gather in big country, of training a horse or a dog or a child, of standing up to the bone-wearying grind required to lean into the wind and care for stock day after day at 50 below. The power of Mary's storytelling, her eye for detail and characterization, and her ability to capture the essence of landscape, weather, and animals bring a freshness that keeps a reader turning page after page. She gives frequent glimpses of the sense of play that can erupt when the work goes well, and of the complex satisfaction that can follow the successful resolution of a day that veered horribly off track.

But most riveting for me are her accounts of the difficulties that families face when they work together. Succession is wrenching, leaving the older generation bereft of identity and purpose, even as younger members chafe to take the bit in their teeth and run. In the chapter titled "Never Meant to Be Unkind," Mary remembers stepping into the kitchen as a small child to find her father and grandmother in a standoff. "We'll be dead pretty soon," her grandmother was saying, almost shaking with rage, "and you can run it however you want. But we're not dead yet." Mary had never seen grownups argue before and it frightened her. She was relieved when her elders returned to their more congenial ways. But in a few years the name on the ranch sign changed from "John C. Budd and Son" to "Budd Hereford Ranch, Incorporated," and her grandfather became the noonday campfire cook instead of the roundup boss. Later she watched the pattern repeat with each generation until she and Stan were the ones struggling to step gracefully aside.

Nor is it easy to work things out among siblings. In "A Ranch Divorce," Mary describes the painful split not of man and wife "but instead, the permanent division of land, livestock—and, too often, a family." Stan and his brother's "divorce" came in the 1970s when the bottom fell out of the cattle market at the same time that interest rates soared. For the Flitners, as for ranching families across the West, debt skyrocketed. The hard times brought a sense of failure and blame. The brothers had worked together harmoniously for years; their kids ran in a pack and their wives were best friends. But increasingly they realized that the only way for either family to survive was to split the ranch.

Interpersonal relations determine the success of business partnerships as well as families. Once on their own, Stan and Mary enter into two collaborations, one lasting ten years, another more than thirty. As each partnership unfolds, we witness how the complicated cast of characters juggles their various personalities, disparate management styles, and skills; how they keep the different leases and ever-changing herds of cattle straight and their kids engaged and working well together; and how they handle the inevitable blowups and misunderstandings in ways that allow them to emerge not only with mutual high regard but also free of debt.

Mary's is a voice that is matter-of-fact but full of heart, that doesn't milk the hard parts for drama but doesn't gloss over them either. Nowhere is this more true than in the glimpses she gives us of what it means to be a woman in such a world. Mary would not describe this as a woman's story so much as a rancher's tale. Nonetheless, she is clear about the particular challenges that come with her gender.

One time, Mary, her friend Mike—Mary Francis—and their junior-high-aged boys were sorting cows and calves when cowboys from another ranch turned up and wanted to use the corral.

The big fellow in his fancy cowboy gear shook his head and suggested we just step aside, and he offered to go ahead and do that sorting for us. . . . He might as well have said, "Bless your heart for trying, though."

He stepped over to take my spot, placing his hand on the gate. I didn't move. "No, we're doing fine," I said. "You'll have to wait." I was polite, I think. We faced each other for a moment—five feet of me, six feet of him not counting his Paul Bond boots with the undersloped heels and his high-crowned Stetson hat. . . . After a few moments, he climbed out over the fence to sit in the pickup with his buddy . . . like he expected to wait all day.

Mary and Mike and the boys finished the sorting in short order. The "real men" in the pickup didn't acknowledge their expertise, of course, but the young men in the corral learned that a clarity of purpose can hold its own against considerable bluster.

In the penultimate chapter, after Stan and Mary signed the papers that allowed them to pass on the ranch to their children, their accountant congratulated them. He had worked with ranchers all his professional life and he noted how rarely he saw land transfer successfully from one generation to another. "More times than not," he told them, "it isn't money that pulls a ranch apart. It's the family dynamics."

In the end, this is the story of successful relationships—within family and also among partners, neighbors, government agents, hunters, environmentalists, and all the other people who intersect in one way or another across the vast open patchwork of private, state, and federal lands that define the ranching West. It is a complicated terrain, and this may well be the first time anyone has mapped it with so much skill.

◇— Preface

For more than one hundred years, ranching has been the center of our families' lives: of my own family, and of my husband's, as we joined the five generations that came before ours into the present day. For my husband and me and our children, success and survival were made easier by the help of friends who came along for the ride—for the adventure and sheer fun of being involved in a rowdy world of ever-changing work and unpredictable challenges. Many individuals enriched our lives with companionship, wisdom, loyalty. And yes, they offered strong backs.

One such friend was a Catholic priest we called Father Pete, who, perhaps, started me toward writing this book. He had grown up as a city boy, but when he arrived in Wyoming, he took quickly and naturally to our West and its customs. He signed up for riding lessons at a nearby equine center, he bought a saddle for himself, and he set about learning livestock and ranch work. When he asked us where to buy a horse, we said, "No reason to do that. We have horses here; they need riding, and you're welcome to ride one of them. You can use Vanilla. She's a good cowhorse, and she'll be yours, whenever you want to come along." He did come along, as often as his parish work allowed. He shied at no job, and we put him through some hellish days of physical toil, but he stuck with us as a friend and cowboy, thick and thin. Father Pete pulled with us, through weddings, baptisms, illnesses, financial challenges, and difficult personal and emotional times that we sought to balance against the burdens of contrary weather and seemingly endless work.

After a particularly exhausting day horseback in the high country—the kind when nothing went right, when the cattle were balky, the sun was hot, and we'd worked ourselves and our horses to a sweaty, plodding fatigue—I thanked him and apologized to him as we rode back toward the corral together. I shook my head, and tried a smile and a weak joke. "What a day. I thought it might even make you take the Lord's name in vain! I don't know why you keep coming. I don't even know why *we* do. It's just ridiculous. Ranching's a disease, I guess. It's an addiction, maybe, a habit I can't break."

Father Pete stopped his horse and turned it toward mine. He took off his wide-brimmed cowboy hat, wiping his forehead as he nodded slowly. He spoke earnestly, then, and frowned a little. "Mary, ranching is a vocation, and you must treat it respectfully. Your family has served this calling for more than a hundred years, and you're a part of that huge, raw story—a grand and powerful story, despite its uncertainties and challenges."

That conversation stayed with me, and I began to consider that "calling" as it affected my family and others. I looked back to review the journals I'd kept as ranch records, through most of my married life. In them, I routinely recorded weather patterns, cattle markets, inventories of hay and grass, and I listed work priorities, problems and solutions to be found, and I named our helpers as they came and went. Family dramas large and small crept in, too, usually given only a few terse lines.

I went on to research old family diaries and historical records, comparing the lives of my husband, myself, and our now-grown children to the lives encompassed within these other writings. I felt great satisfaction when I found factual references to familiar geographies and personalities. I realized that perhaps my children or grandchildren would someday wish for similar observations, seeking answers to their own questions, yet unasked.

I began to realize the importance of explaining our "vocation" to the generations who will come after us, by writing how and why we're still here, working on the land. I felt a calling to describe ranch life with authority and honesty, and to offer insight into the skills, culture, and knowledge held by those within this profession. The logic and lore of ranch work take some explanation, as do our unspoken philosophies; my journals offered a rich foundation for such telling.

I cherish the magical visions of ranching found amid the myths and romantic images of old-time cowboy ways—tales that are funny or sad, or filled with bravado or trepidation. An honest retelling, however, acknowledges resilience and commitment, balanced by luck, timing, derring-do, and, inevitably, hard work. Disappointment, sorrow, and frustration come, too, alongside joy, happiness, and satisfaction.

What came forth is this collection of stories, describing the realities of a lifestyle that I loved at childhood and have ever since—as wife, mother, grandmother, and ranch partner. The stories are not chronologically presented but are instead written to reflect the patterns that repeat within ranch settings and elsewhere. I discovered the "addiction" to be a search for the peace that will come when the work is all done, when the grass is green, when the fences are tight and the cattle are fat; the satisfaction of knowing that the range is ready, the horses are ready to ride, and the hay is in the stack. That peace doesn't last—things are never "just right" for long, and whenever things aren't "just right," we'll try again. It's the habit we can't break.

My Ranch, Too

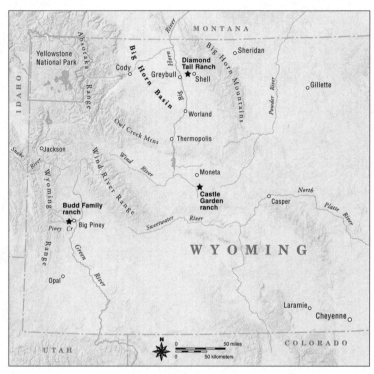

Wyoming, with Budd, Diamond Tail, and Castle Garden ranches. *Map by Ben Pease.*

1

Firecrackers, 50 Cents

In 1879, a Kansas man, Daniel Bockins Budd, received word that his younger brother Philip had died suddenly in a small northern California town close to the Nevada border. A few days later, Daniel Budd left his small farm, his job at the prison in Atchison, his wife, Josephine, and their four small children, and traveled by train to Eureka, Nevada, to settle his brother's final accounts.

Daniel Bockins Budd was my great-grandfather, the patriarch of the Budd family members who live in Wyoming today. His small, hand-written diary recorded the monumental change of direction his life took after he heard of his brother's death. This journal leaves questions among the facts, but it provides the basis for telling and retelling an enthralling frontier-day saga that began with a few lines, penciled into his life's record.

An area newspaper gave notice of Philip's death, but it also indicated some confusion at his passing:

BUDD, PHILIP P.
Placer Weekly Argus, Auburn, Saturday, 2–15–1879

Sudden Death at Borland's Hotel—A man named P. P. Budd, a traveler registered at Borland's as belonging to Eureka, Nevada, was yesterday morning found dead in his room. He had arrived only the evening before from Stockton, so we learn, and he seemed to be a respectable-looking miner. He had in his possession $605 in money and a note for $1200. His death is believed to have been caused by disease of the heart. He was about 35 years of age. The news of his death was telegraphed to Stockton where it was thought his friends lived. But the answer came that, so far as known, he had no relatives there. The Odd Fellows of Austin, Nevada, telegraphed to the Lodge here to take charge of the remains, which was done.

When Daniel Budd began his journey, on February 19, 1879, he wrote in his diary, "Started from Atchison, Kansas with three hundred and eighty three dollars." The diary makes numerous references to railroad fare or "R Road fair," as Budd wrote it, and he apparently traveled directly to Eureka, Nevada, where he learned that his younger brother owned a sizeable number of cattle and horses. No facts were recorded, or at least did not survive, to explain how Philip, at age thirty-five, had acquired these assets. My attempts to retrace Philip Budd's presence in Auburn indicate that he had gone to take the mineral baths there, said to be medicinal—and that he had died of natural causes, perhaps the "disease of the heart" as described in the newspaper. A white rock headstone bearing his name stands in the Auburn cemetery, marking his life and death.

Budd's diary goes on to record a three-day train ride to Eureka, where he found a hotel, paid a dollar for a room, and, as his diary says, "wrote to wife." The next morning he "came out on ranch 45 miles" in an area known locally as Grubs Well. The diary doesn't say how he traveled the forty-five miles (a considerable distance), whether by wagon or saddle horse, to locate his brother's property. And it doesn't say who showed him the whereabouts of the cattle, although a March entry—"tired you bet"— suggests a long day's viewing.

After checking into the Eureka hotel, where he headquartered for the duration of his visit, Budd undertook to gather the cattle and to count and brand them, after which they could be sold. The cattle market would be more favorable at a railhead nearer the East Coast, he thought, and he set about organizing a roundup and then a trail drive.

A month later, on March 10, he noted, "Eighty dollars for attorney fees and .75 for dinner, .50 for breakfast and .25 for cup of coffee." He paid for washing, tobacco, and "$5 to the Destitute." Three days later, "Visited my Brother's Grave. Wrote to Wife."

The following week his entry was similar, perhaps indicating that he had begun to feel somewhat settled: "Wrote letter to the children, letter from wife. Shaved and Boots Blacked .50 Segars .50 Went to Church .50." The next day, again: "Wrote to Wife . . . Tobaco .25 Beer .10 Bed .50."

These entries suggest that Daniel Budd was an honorable man, since he left more of his money with the church and the "Destitute" than he spent on beer and cigars! Eighty dollars for attorney fees is significant, and that payment must have made a good dent in the small bankroll he'd brought from Kansas.

On March 6, the diary had only one line: "McKay wants cattle." Hugh McKay was a Nevada man, presumably an acquaintance of Philip's, who would become Budd's business partner. On April 13, Budd "wrote to McKay." A month later, the two men signed a formal agreement duly recorded in Eureka County, Nevada, May of 1879. The agreement specified that "as soon as a favorable opportunity offers, and a sufficient amount of money can be realized to warrant a reasonable profit, the two parties will sell the same at any point on their route easterly."

The partners agreed to use Daniel Budd's brand, the 67, and they proceeded to find a crew of cowboys and then "commence," as the diary said, to round up and brand the cattle, bunching the cattle where they found them on the unfenced Nevada rangeland. Budd referred numerous times in his diary to this "rodear," a commonly used Spanish word that means "to encircle" or "surround."

He used various spellings in his diary—"rodair," "roder," "rodiar," "roduir," and others. (The word functions interchangeably as a noun or verb, and is used among some cattlemen in present times to mean "roundup," or a gathering of cattle.)

When they had the cattle together and branded, the partners still faced a daunting journey from western Nevada to a point of sale, probably Nebraska, as the diary later indicates. Budd's diary follows this undertaking. To Omaha, it would be approximately fifteen hundred miles, one day at a time. Only in a stretch of the imagination could a reader glean in his account the "life of the cowboy" as it is romanticized in modern-day movies and television shows. What appears as colorful and adventuresome in these dramas doesn't sound so entertaining in Budd's diaries. He recorded mundane details nearly each day, commenting frequently about the weather as he and the cowboys worked their way through the big country. "Snowed some in mountains and Rained," and, "I most froze." His diary names some of the men who helped him find cattle in an area he referred to as Empire Ranch. Here he had "jerked beef and dry bread for diner and super," and remarked, "Would like a change." On June 5, 1879, he "Bought 1 Bay Horse $50.00, suit of Close $19.00," and then "started on Rodair." That day he "rode 40 miles, got snowed on, had supper at Willows," where he stayed the night and paid $3.50 for hay for the horses. A few days later he wrote, "40 miles from Eureka and lost in mountains."

He offered no detail when he wrote, "visited by Nevada Ladies at Empire Ranch."

Even in early summer, the weather was brutal. "Raining and Snowing. Put pockets in my pants. Joined the rodiar Party, Rained Snowed, got cold & wet, thundered heavy." That same week, "My horse fell with Sam Lowks. I thought it had killed him. Slept under a big rock."

The cold weather continued. On June 17, Budd wrote that he "Bought a horse of Dic Fagen for 70 Dollars," and, "Bought 1 Pr

blankets $10.00." "Cold Ice ½ inch thick Slept on ground. Held
Pravo all day with overcoat on." He continued, "on the Rodair
Branded 13 Calves. Got Kicked by a mustang horse."

Apparently the Fourth of July brought him back to town for the
holiday, where he bought another horse for $45.00, had it "shawed"
for $3.00, plus "To Hors Shoes $15.00," and then he treated himself
to some celebration:

July 3, 1879: "Eureka Breakfast .50 Went to the Races 1.50 Post-
age Stamps & Box Rent 2.00. Fire Crackers .50"

Firecrackers? When I read this part of Budd's journal, I have
trouble seeing this fun-loving, jaunty cowboy as the same man
who had recently uprooted his life and rededicated his days to the
hard work and purpose of this enormous roundup and then the
trail drive. Here he now was, visiting the horse races and buying
firecrackers! Only this one reference hints at a rowdy, jolly man,
if indeed he was—or reveals confidence and bravado and daring.
Yet that side of him must have been there, setting him up for the
adventures and risks he faced.

We can guess that Budd had used up all the firecrackers and
had had enough fun, when a few days later he returned to "Roder,"
continuing with his cowboys to search for his brother's cattle,
branding small bunches here and there as they located them. They
worked on through the rest of July, as entries show: "got no Cattle."
"Am getting mighty tired." "Rode 45 Miles."

Finally, July 13, a change from dry bread and jerky: "Catched
Sage Hen Rosted it and eat it."

And then, "Old CharleyHors & colt got mired." Typical of his
terse, brief entries, Budd didn't write down how this took place, or
whose horse Charley was, or how it all turned out. As with many
diaries and other daily accounts, the obvious often goes unnoted.

July 28, 1879: "Hired an Indian to Drive Horses to camp. Cattle got
away at night. Stood guard from 10 to 3 tired and Sleepy." It would
have taken time for newly gathered cattle to become accustomed to

staying in a big, loose bunch, but eventually they probably adjusted to a routine of sorts, grazing and resting.

August 4, 1879: "on Roder . . . sorted Cattle & branded," and, "Indian Bill for Riding Colt $2.50"

Finally, three months after Budd's arrival in Nevada, the trail drive began.

———

August 10, 1879, "started for
Newbraska with 777 head of cattle."

If we made a movie about Budd's forthcoming venture, it would show a cook and a chuck wagon with a team of horses to pull it against the huge backdrop of sprawling landscape—cattle milling, horses galloping, spurs jingling as the cowboys set their hats and hearts toward a new country. The cowboys would appear in full persona; we would know how each looked, what they wore, and how they talked. Budd never mentioned personalities, though, or who he liked or didn't like. Old photographs show him as a handsome, strong-featured, unsmiling man, with dark hair and a heavy black beard. He wore a brimmed hat, knee-high boots, and a vest and a neckerchief, but other than the occasional shave or haircut, and a new set of "close," his diary doesn't waste words on appearance.

The "shoot 'em up" western novelist Louis L'Amour would have found a moral value and a code of ethics in Budd's story, or at least a sense of honor. If Budd were cast as John Wayne in an old-fashioned Western, he would have commiserated with his companions, giving us their dialogue in colorful fragments of sentiment and compassion. No such expression exists within Budd's diaries—no motivational language, no inspiring, hopeful pronouncements.

Budd makes no mention of six-guns or weapons, although probably he or his cowboys carried some rifles in their wagon, if only for hunting. It seems the only excitement by explosion was that of the Fourth of July firecrackers. In fact, the roasted fool hen was "catched," not shot, and he doesn't say how they "catched" it. I'd like to think that event called for some levity, a novelty to the cowboys, since he made a point of mentioning it in the diary.

I could not tell from Budd's account how many cowboys began or finished the trail drive together, or how old they were, or where they came from. He noted only a few cowboys by name, and he didn't record what he paid the men or what their qualifications or skills were. The diary vaguely mentions some he hired and some who quit—and some who wanted to!

Thus, through August, the cowboys moved their herd along a route that would provide water and grass, and much of this trail can still be traced on topographical maps of Nevada, southern Idaho, and finally into Wyoming.

At the beginning of the route, they "Campt all night at Hay Stack, Damn Poor Feed" to "Rabbit Creek 10 mile pretty good feed," and then moved on to "Lamorl Creek," which I believe is near Lamoille, Nevada.

The diary tells us they "camped on Lamorl Creek, Feed Short, plenty Water, Lost 7 head of horses one Big Steer Died of Poison on Lamorl."

"Hunted Horses all day, found them at night." The trail drive moved on to Humbolt, Nevada, crossed Mary's River, up Emigrant Canyon, and continued in a circuitous route toward southwest Wyoming Territory, from where the men hoped to head farther east for the railheads and shipping yards in Nebraska.

In late August, Budd "Mailed a letter to Wife from Deeth." There are numerous mentions in the diary of letters between Budd and his wife, but none of those letters survived through time. I assume he was a man of few words in his letters, too, and that his correspondence with his wife simply followed the telling of his journal: conditions remained difficult, and the weather cold. "Horse fell not Hurt. Ice ½ inch thick," his diary says. On August 29, 1879, the diary tells that the cowboys "Moved camp to Humbolt Wells. Bill Blakr, Sam Louks, Crazy Bill & Lawrence Dogie wanted to quit." At one point he mentioned sending 1.50 to "little Charlie," one of his sons at home.

On September 3, 1879, he wrote in the diary that "the cook got lost. Dry camp got to camp 11 o'clock p.m." Probably the cook had gone ahead with his wagon to set the evening's camp. Since the camp was not where the cowboys expected it to be, someone must have searched for it, but this one line is all Budd wrote of the incident.

The arduous journey appears to have averaged five to ten miles each day, as nearly as the map can be followed. Budd referred to areas of "good feed, no Water," and "pretty good feed," and then "plenty feed, terrible hard driving" as the cattle grazed, determined to fill their bellies, reluctant to move. A week later, he wrote, "tired hungry & Dirty Renshed my socks."

In mid-September the drive reached Raft River and the border of Idaho Territory, and then continued through southern Idaho. Geographical references to this part of the route include Soda Springs, Montpelier, and the Thomas Fork of the Bear River.

Things got no easier: "Laid over at twin Springs. Changed close, found graybacks plenty, one cow Died of Poison."

On September 20, 1879, the cowboys "Moved camp 12 miles over Banack Mountains. Hardest Days work on trip. Got to water at Dark. Supper at ten oclock. Camped on Malad Creek."

By then, Budd and the cowboys had trouble keeping the cattle from drifting out, probably in search of feed. There are several references to night herding and standing guard at night to prevent cattle from going back. Nearly every day, the diary mentions whether they found feed or water.

In Idaho Territory, small communities and settlements had been established, and so Budd was able to buy supplies: "green corn & Potatoes & Butter," and "a cheese 20 cents per lb, 21 lb of Flour 5 cents per lb."

He took care of other business as well: "Sold 13 poor cows to John S Watson at $10 per head. Mailed a letter to wife." Some of the cattle were showing the stress of the miles behind them, "poor" meaning thin or lame, and rather than take the financial loss if they were to die along the trail, Budd sold them.

Late September: "Passed through Soda Springs. Bought myself pr Boots $4.00 Bought myself Hat $2.00," and "Snowing like hell."

I don't know exactly where the trail herd was when Budd wrote, "Laying over. Raining and Snowing hard very Disagreeable Mormons plenty," but it appears it had reached the lower forks of the Green River.

Budd and McKay had planned to winter the cattle at Fort Bridger, Wyoming Territory, then continue the last 650 miles to the Nebraska railhead in the spring. Unfortunately, when they reached Fort Bridger in October, they learned that Mormon settlers there controlled that broad landscape as well as that to the south and west, into Utah and Idaho. "First come, first served," as they said. The Mormons refused to let the large 67 herd graze on the winter range, likely saying they needed the land and grass for their own livestock. The Mormons advised Budd and McKay to go on "up the Green" to find winter range at LaBarge Creek.

I can only guess at the anxiety Budd and McKay felt as winter neared. In his diary Budd wrote, "has the appearance of More Snow. Cannot say where we will Winter yet. Cattle are doing pretty well considering the Storm on herd."

October 21: "Laying over on Hamsfork of Green River. McKay started to Fontnell to look for winter range. Splendid weather."

Budd "laid in camp," he said, waiting for McKay to return. "Plenty of deer and antelope in hills." And, a few days later, "Laying on Hamsfork—getting tired—would rather be moving. Have the Rhematis, some in right leg. Rather cool to be comfortable."

A week later he noted, "McKay came back." Together they "concluded to winter on Blacks Fork," another tributary of the Green.

And then, October 29, "Left Hams for the little Muddy 18 miles. No water. Mired 60 head, 10 head Died in Mud . . . Worked all night."

I don't think those of us who handle cattle today—using four-wheel-drive pickups, stock trailers, or cattle-freight semitrucks, following interstates and highways—can really appreciate a day and a night like that one. Budd did not express anger, disappointment, or fear in his words, only a factual accounting of what could have been defeat. The cowboys would have used ropes and horses, or worked afoot to pull, drag, or push the thirsty cattle from the mud, cattle that had waded into a seeping pond or boggy streambed seeking water. The cowboys must have been exhausted, wet, and muddy themselves. When they succeeded in getting the cattle onto dry footing, the men would have had to force them away from the dangerous wetlands. The next day, October 30, Budd only wrote, "Moved cattle from Little Muddy to Blacks fork 15 miles."

By the first of November, the drive had reached some small settlements in this part of the Green River Valley. Budd wrote, "Granger Station would like to go to church but have to herd cattle." Granger, where the Ham's Fork joins the Black's Fork of the Green, had once been a Pony Express stop and then a stop for the Overland Stage, and it boasted a railroad station, but there is no mention in the diary of loading the cattle onto a train there. Perhaps the cattle were too thin to make a profit, or perhaps there were not adequate loading facilities. At any rate, the herd moved on past.

A few days later: "team got Stoled supper at 12 PM." Budd never refers to finding the team, and we do not know what became of the horse thieves. Perhaps the drovers simply continued on, two horses short, but we will never know. Typically, the facts most obvious to the man who was there did not seem worth writing about. After all, the journal wasn't a storybook: it was a business account.

The men continued up the Green River, "crossed the River twice no feed," to the Fontanelle and then LaBarge Creek. There, too, they were told to "keep going up the Green." Their cattle were suffering from the long journey, becoming weak and thin after so many miles. Early November: "left 2 calves 4 head of cows gave out."

———

November 13, 1879: "on Big Piney . . .
up Green River . . . Winter quarters."

Three months after they had left Nevada, Budd noted that they "Counted cattle turned them loos." Although they feared they might not see a cow alive in the spring because of the cold and snow, Budd and McKay made the decision—choices limited by the fast-approaching winter— to leave the cattle in an area where there appeared to be adequate winter feed.

They located near the confluence of several Piney Creeks, as indicated in the diary: "On Big Piney." The partners had determined that McKay and one hired hand would remain with the cattle throughout the winter, when the "boys hauled wood," and "hauled poles for corral." They built a cabin, too, and a note explains that McKay had taken a wagon to Green River City to stock up on supplies. On November 20, Budd wrote that he "Run some cows up Little Piney beautiful day, washed, more gray backs." When McKay returned, Budd and the rest of the cowhands drove the herd of extra horses 250 miles back to Salt Lake City, Utah, where Budd caught a train home to reunite with his family in Kansas.

I don't know how Budd's wife, Josephine, had managed their family's daily needs while her husband was away, or what her own hardships were. Daniel Budd was forty-two years old at the outset of this journey, and perhaps he and his wife had determined to seize this opportunity as it offered itself, to make necessary sacrifices for the sake of their family's future. Budd's references to their correspondence, intermittent and sparse as it had to be, indicates a mutual commitment to some success with the cattle drive.

Budd returned to North Piney Creek the next spring, where he saw that the cattle had survived the winter in fine shape. As McKay had learned in his time there, plenty of land was available for settling; only two other ranchers (Otto Leifer and E. Z. Swan) had located in the area. Budd and McKay decided to settle there, rather than continue on east, and thus was founded the Budd ranch in Wyoming.

Budd sent for Josephine and their four small children: Sarah (Sadie), Charles, Jesse, and John, and the family declared Big Piney, at the confluence of three creeks—North Piney, Middle Piney, and South Piney—to be their home. Two more sons, Henry and Daniel, Jr., were born in Wyoming Territory in 1880 and 1882 respectively.

In 1885, Budd sold his share of the 67 to Hugh McKay. Budd seems to have been a businessman at heart, restlessly looking for new opportunities and challenges. He established a new headquarters, which became the present-day town of Big Piney. He built a home, a blacksmith shop, and a large two-story log building that became a community store, post office, dance hall, and boarding house. He became a prominent figure in the region, an icon of perseverance and prosperity, among other things donating land to the Big Piney community for a school.

Daniel Budd died suddenly of appendicitis at the age of sixty-two. Josephine lived to be 102 years old, living most of her adult life in the home her husband had built for them.

A story my father later told indicated that my great-grandfather Budd never forgot the way he and his drovers were treated by the Mormons at the time he entered Wyoming Territory. When my dad left Big Piney in the 1930s, to attend college in Logan, Utah, he looked for a room to rent in a boarding house there. He described how he went to answer a newspaper ad at an address in Logan, where he was greeted by an old couple rocking in their chairs on the front porch. Dad explained why he was there, and thought he had a sure thing when the man said, "Your name is Budd. Any relation to the Budds in Big Piney?" Dad replied with enthusiasm, "Oh yes, that's my family." The man said to him, "Well, let me tell you. Back when my wife and I were young, we drove a team and wagon past that store Daniel Budd had there. We were tired and hungry, and we said we wanted supplies and maybe a room. Budd said, 'Are you Mormons?' We told him yes, we were Mormons, and Budd just said, 'Well then, keep on going.' Dad said the old man stood, pronouncing, "So now I can say to you, keep on going."

Daniel and Josephine's five sons and only daughter remained in the Big Piney area, and it appears that they, too, were independent and ambitious. The son Charlie established his own little town, building a bank and store on the hill immediately overlooking Big Piney. Jesse continued as store owner and postmaster, the latter a position he held for forty-five years. John (my grandfather), Henry, and Daniel, Jr., became entrepreneurs and stockmen in their own right, while promoting community expansion, irrigation projects, and oil development. Daughter Sarah, "Aunt Sadie," married cow-hand Al Osterhout, and they put together a large ranch. Aunt Sadie was said to be a sharp, intimidating person; although small in stature, people often said she "wore the pants in the family." I asked my father once how that could be possible, since women were not well accepted as business owners or ranchers back in that day. Dad replied, "As far as I know, it worked fine. Al ran the ranch and Sadie ran Al."

John Budd married in 1905, having accumulated land and cattle for himself and his wife, Lula McGinnis. They passed that ranch on to my parents, Joe and Ruth (Peterson) Budd, and it remains intact, owned by my younger sister, Nancy, and her sons. Not an acre has been sold.

My older sister, Betty, married into a local ranch family by the name of Fear. The Fears arrived in Big Piney in the late 1890s and registered their brand at that time. The Fear family has a wealth of stories of their own to tell, and their property remains within family ownership.

My own life took me away from Big Piney when I married my husband, Stan Flitner. But marriage did not take me away from ranching. My husband's grandfather, Arthur Flitner, had purchased a ranch in northwest Wyoming in 1906, which our family still owns and operates, and where Stan and I live. None of the original holding has been sold.

Family stories and ranch-life stories connect, winding back and forth throughout Wyoming history and tradition. My great-grandfather's diary is the foundation for the telling of these and other stories, and it is a privilege to share them.

2

Never Meant to Be Unkind

The phone would ring at the same time each morning, on the old party line at the ranch. Mom would roll her eyes. "It's *Mother*," she'd say. She'd take a long drag on her cigarette, shake her head, and pick up the receiver. "Mother" was my grandmother, Lula McGinnis Budd, Dad's mother, the ranch matriarch. "I don't know why she calls," Mom would say. "There's nothing to talk about. Always the same." She'd sigh. It was the 1950s, so no caller ID of course, no answering machine. "Hell-o." And yes, it would be Grandma, and yes, she called every morning.

Mom was impatient with her, but I saw my grandma as busy and jolly, always with a ready smile. Grandma had earned respect among her ranching neighbors; she was knowledgeable, practical, and savvy with livestock, too. When I was a little girl, I accidentally rode my horse Trixie into a tangle of barbed wire behind Grandma's house. I started to cry, terrified that Trixie would spook, that I'd fall off, or that the horse would cut her legs. Grandma saw us from the clothesline where she was hanging laundry. "Be still. Just sit still," she called. "Don't move." Wearing a cotton dress as usual, and her sensible shoes and stockings, she carefully stepped near, calmed the horse and me, and capably bent down to free the horse's foot. I idolized Grandma, that day and every day.

A different memory: Grandma standing toe-to-toe with my dad in her kitchen on a hot summer afternoon after we'd returned from a long day of riding. "Stay for a bite to eat before you leave," Grandma said to us. We kids sprawled across the living room floor, tired and hot. Grandma wore her housedress and apron, having changed out of her cowboy clothes, and she had her white hair tucked under a hairnet. Suddenly the mood changed, and I didn't know why. Short and sturdy, with her hands on her hips and her Irish blue eyes peering through her thick glasses, Grandma was shaking mad. Dad seemed shocked. "Well," she said, her jaw set, and her voice hard, "we'll be dead pretty soon and you can run it however you want. But we're not dead yet."

Their confrontation scared me; I'd never seen them angry at each other, and I thought grown-ups didn't argue, because they usually didn't at our house, at least not in front of the kids. I don't know what the specific issue was or how they settled it. I remember Dad walked out and slammed the door, but they seemed to behave cordially when I next saw them together.

Now I know that generational transitions from boy to boss, boss to bystander, and, finally, from heart to heart are precarious. I hear echoes of Dad and Grandma in my own family's conversations. When my grown son suggested I move away from a job in the corral—one I'd done for years—to let a younger person take over, I turned to face him, hurt and angry. "What do you mean, that will be easier for me? Don't you think I can decide that?" I snapped.

"Relax, Mom. No big deal," he replied. He was trying to be considerate; I was annoyed and embarrassed. Others were listening. I wanted to avoid a scene, so I stepped aside, yielding my spot. These awkward moments pass, but I always think of Grandma: "I'm not dead yet."

Grandma and Grandpa Budd married in 1905 and homesteaded in southwest Wyoming, near family holdings along the Green River. As ranchers, they survived two world wars and the Great Depression, and more than one terrible winter and drought, barely holding on to their land and cattle. They established a successful ranch that is still in the Budd family a century later. Their commitment to each other triangulated exactly with their love for the ranch.

My dad was their only son, handsome heir apparent—smart, ambitious, and probably overindulged. When he returned to the ranch after college and then married, Dad and his parents squirmed in their changing roles. In a family scrapbook, a handwritten letter, signed by Grandpa Budd in 1946, shows the stately, beautiful letterhead used on behalf of "Meadow Canyon Ranch, John C. Budd and Son." The ranch signpost of my childhood memory used that name, too, but by the 1960s, the sign said "Budd Hereford Ranch, Incorporated." It is difficult to imagine how or when the two men agreed to pass that torch of responsibility and stature. I never knew.

At some point men began to work for Dad, though, not for Grandpa and Grandma. Grandpa was a top cowboy and a capable leader, well known in the ranching profession. Horseback, he made a distinctive impression—he wore a short-brimmed hat and sometimes bib overalls or loose-fitting chaps. He wore spurs on his lace-up work shoes. I didn't notice when he became the noonday campfire cook at the Deer Hill Roundup instead of Roundup Boss, the job he had traditionally held. A stiff knee required he walk with a cane; gradually it became difficult for him to ride horseback at all. He disappeared from the hay crew, too, and only showed up bouncing the car out to the field sometimes, bringing iced lemonade.

Dad bought more land, created a prestigious herd of registered Hereford cattle, and became involved in off-ranch politics and activities, asking Grandpa's advice but only infrequently. As time passed, Grandpa and Mom carefully juggled the ranch bookkeeping job between them. Mom paid the bills, after which Grandpa hand-entered every transaction into a lined ledger, scrutinizing each expenditure, frowning and puffing on his cigar.

Rather than tend chickens or spray weeds to "keep busy," Grandpa and Grandma moved to town. They didn't want to be "underfoot," they said. They bought a big new Buick Roadmaster and gamely traveled to Mexico and California and Arizona. When they came back home, town life and retirement bored them.

Yes, Grandma phoned every day. Grandpa was too proud to call, but he, too, was curious about the "goings-on" at the ranch. Together in the Buick, they drove slowly along North Piney Creek nearly every day "to see what we can see," Grandpa said.

Grandma died at age eighty-three and Grandpa at eighty-nine. Through their last days they loved the place they'd built, perhaps imagining they could protect it with their caring vigilance.

Grandma's angry words on that day I remember from childhood call to mind that time that inevitably comes to a business or a family, when an older generation steps aside. I know Dad respected and loved Grandma and Grandpa, and I know he and my mother never meant to be unkind.

Eventually Dad and Mom reached that same plateau: my sister and her husband returned from college and the army, excited and ready to begin a life in ranching. Dad handed off the management to them quickly, sternly insisting, "They need to have the responsibility and make the decisions." Dad and Mom moved to town, to the same house where Grandma and Grandpa had lived.

At first they drove to the ranch frequently to help with work projects. As time passed, Mom phoned my sister daily. She kept busy at home babysitting and baking for the grandkids, peering out the window, hoping someone might stop by. If no one showed up, she happily called to offer help with the laundry or any odd job, as she always wished to be useful wherever she was.

Dad thought he would like retirement since his strength and enthusiasm for ranching were diminishing, but he soon became grumpy and disapproving. He puttered in Grandpa's old wood shop; he glumly watched television news. As Grandpa had done, he

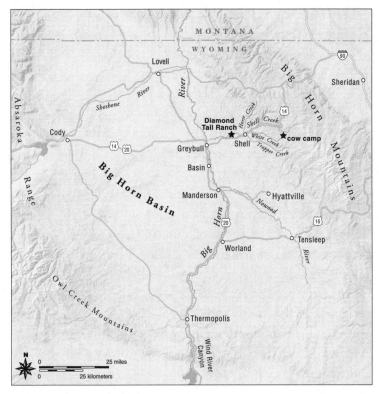

Big Horn Basin. *Map by Ben Pease.*

trudged resolutely to the post office each morning. He rejected our well-meaning suggestions about finding hobbies and new interests, saying, "Wait and see. I've been your age. You haven't been mine."

Three hundred miles north, in Wyoming's Shell Valley, my husband's family took a similar course. Arthur Flitner, that family's patriarch, had come to Wyoming after Oklahoma's 1889 land rush saying later that he was "not fast enough to do any good." He was said to be feisty and impulsive, a shrewd businessman who built his cattle ranch in the Big Horn Mountains during the early 1900s with money he made in a lumberyard and grain elevator in Norman, Oklahoma.

In 1929, Arthur's only son, Howard, married his true love, Maureen, in the familiar western pattern—"rancher marries schoolteacher." Arthur passed the ranch to Howard, and family stories describe their confrontation when Howard bought a band of sheep without consulting Arthur. At the corral, Arthur angrily challenged Howard: "What are you trying to do, break the outfit?"

Arthur and his wife, Anna, moved into town; they eventually moved to Fromberg, Montana, and then back, while Howard and Maureen, my husband's parents, dabbled in dude ranching and brought new ideas to the family ranch. The sheep stayed.

In turn, Stan and I married and returned to that family ranch in Shell Valley. Ranching had hit difficult economic times then, in the 1960s, but we were optimistic, excited for our future. Howard was in his sixties; he seemed ready to delegate the worry and responsibility to the younger generation, and we assumed he was relieved to be free of the heavy work. Maybe, though, we just elbowed Howard and Maureen out of the way; after all, we were young and bold, if well intentioned. They built a new house on the property, and Howard put in a garden and kept his old saddle horse, Apollo, in a little pasture nearby. Nearly every day, Howard drove over to check on things; we shared coffee often. Maureen was a jolly, loving grandmother to our children and appeared to keep busy enough with card clubs and social activities.

They seldom called to ask for favors, telling us later, "We didn't want to bother you." Stan left Howard waiting for him at home one day when they'd planned a work project together. When Stan came home that night, Howard phoned. "I waited all day," he said. "I don't ask much. I try not to bother you. But I think I deserve some consideration."

"I was in a hurry, and busy," Stan admitted, "and I just forgot. I'm sorry. I'm really sorry." For Howard, the apology did not erase his disappointment and humiliation at being so useless as to be forgotten.

In those same years, I chatted with a ranch woman in her forties, about my age, I guessed, as we sat next to each other at a cattle auction in Montana. Although we had never met before, Bonnie and I comfortably shared our stories while we waited for the sale's end. She told me about their partners, her husband's parents, who spent the winters in the warm south. "Bill's dad phones every *day*," she said, smiling. "We fax all the accounting to Arizona, can you imagine?

"Honestly," Bonnie continued, shaking her head. "What a pain. Wouldn't you think they'd play golf or swim or something, and enjoy their retirement?" We laughed and talked light-heartedly about how *we'd* enjoy life when it was our turn to step down.

Two decades later, it *was* our turn—and it took me by surprise when I realized how quickly time had passed. In 2004, Stan and I built a new home a couple of miles away from ranch headquarters, letting our younger generation replace us in the old ranch house near the corrals. "We're retired. Picking our jobs," we said. "Taking it easy." Now, in our new role, we make jokes about being put out to pasture, but they're not very funny.

Friends our age share our uncertainty and fear of being without purpose, without respect. A ninety-year-old woman rancher angrily expressed her bitterness to me, saying, "Can you imagine? They told me to stay out of the corral, that I'd get hurt. And it's my corral!"

"I'm just so damned clumsy," my neighbor says, rubbing his gnarled hands together. "The 'kids' and the grandkids say they'll help me anytime, but I have to ask, and then probably ask again. They act like they invented 'busy.' Mostly I hate to *need* help."

"I know I ain't all that speedy anymore," a friend tells me. "It's bad enough, being stiff and slow, but it ain't like I don't know what I'm doing. Who do they think got 'em here, anyway? I let it go, but when she told me they didn't need me, I was steamed. She took that branding iron right out of my hands."

"Yeah," I say. "Let it go. They don't understand. But they will, someday."

The reality should be no surprise—ranching is not a lifestyle that accommodates old age or frailty.

I walk out to the mailbox each day as Grandpa and Dad did, looking for purposeful activity and conversations. I drive to the corral to "check on things" that don't need to be checked, and I ask questions about things I don't need to know. I smile when I see things well done, and I enjoy my view of a beautiful ranch.

I phone too often, perhaps, and I imagine our sons and daughters rolling their eyes, saying "It's Mother. *Again.*" "You didn't call," I say. "We could have come yesterday, had we known."

"We were busy," they reply. "Sorry. We forgot to call you." I know I speak sharply sometimes to hide my hurt at any slight, preferring anger to pity.

Some days I feel the pinch of my grandma's shoes, or my mom's, remembering their sighs and frustration as the years overtook them. I thought I'd be more graceful, more accepting of change. I thought the hand-off of responsibility and authority would be more reciprocal, and more gradual. I don't like thinking of myself as angry, or sad, or fearful. Surely I know growing old is inevitable.

Stan and I talked last night, sheepishly referring to a petty slight the day before. He swirled ice cubes in a glass, saying thoughtfully, "It's just what comes. It's the way it is. No harm intended yesterday. Let it go. We have to let things go."

"Wait and see," as my dad said. "I've been your age. You haven't been mine."

I'll let it go. I never mean to be unkind. And I believe no one else does, either.

3

Little Bitty Cowgirls

The ranch hands rolled their evening smokes outside the bunk-house as they sat talking. They didn't see me walking past, and it surprised me when I realized they were talking about my sisters and me. "That seems like a helluva bunch of cows to gather tomorrow, with just those little bitty girls for help," one of the new men said. He continued, striking a match, "Don't see how we'll get it done, all those miles to go." He took a long draw on his cigarette, blew out the flame. "Probably just a damned nuisance, babysitting them. Gotta put up with the boss's kids, I guess." Embarrassed, I ducked back into the shadow and waited to hear what was said next. "Well," replied Walt, who had worked for Dad for many years, "we'll get it done. Don't you worry about those little *girls.* They are pretty damned-good hands, and they can keep up. Just see that you do."

Betty, Nancy, and I were the "boss's kids"—three years apart, with blue eyes, broom-straw hair, freckles, ill-fitting clothes, skinny, and little for our age—and on this day, Betty was twelve or so, Nancy about seven, and I, in the middle, about ten.

On those days in the 1940s and 1950s, most ranch kids worked with their families just as we did. It seemed unremarkable that we were *girls,* or young, no matter what the ranch cowboy said. Horseback, we were expected to keep up and do a day's work. We could ride with the other cowboys at a hard trot for miles before we reached the beginning of the work; there were no horse trailers then, no roads through the sagebrush. We did as we were told, complaining among ourselves if we got a side-ache or chafed knees. Mostly we tried not to attract any attention from our grandpa, a stern and impatient man.

On any day, Dad and Grandpa rode out front together and, according to old-time cowboy etiquette, the rest of us followed respectfully behind them as they talked of plans for the day or considered the range grass or exchanged local news. Kids rode last, of course, giggling and teasing each other until Dad or Grandpa scolded us. "Settle down. You'll be worn out before we ever find a cow." Dad helped us if we needed to tighten a cinch or fasten a snap, or sometimes he boosted us onto our horses. Then he'd remind us, "You need to figure it out yourself. I won't always be there." He showed us how to use a rope to open or shut wire gates and how to knot our bridle reins if we had trouble holding onto them.

We girls dressed like the other cowboys, in long-sleeve shirts, hand-me-down denim pants, boots, and brimmed hats, ours tied on with a string. We wore chaps, too—they came from F.M. Light and Sons, the historic traveling mercantile of Steamboat Springs, Colorado. The older Mr. Light drove a regular route, selling his way north through ranch country as far as Jackson Hole, and Dad ordered the chaps custom-made for us—three pairs in stair-step sizes, tiny batwings of rough-out leather. When Mr. Light unpacked the chaps from his hearse-like van, he showed off the nickel conchos and the snap-shut pockets. "You can carry some hard candy, see? Or your gloves," he explained. Dad agreed, beaming, "Maybe your legs won't get so sore on the long rides. Chaps are warm, too, when you're horseback."

Betty, Nancy, and I rode the horses Dad chose for us. Each of us started out with gentle old Brown Jug and graduated up to Smokey; as time passed, Dad found horses that matched our advancing skills. When I was about ten, Dad found a little brown mare for me. I wanted to name her Thunderbolt or something glamorous, even though she had no flashy markings, no white stocking legs, no starred forehead. But Dad said, "Her name is Trixie."

Trixie was a three-year-old and only green-broke, but she was gentle and smart, and Dad thought I could manage her. When spring came that first year, though, Trixie began to "feel her oats." I only weighed about fifty pounds, and she bucked me off a few times—not with bucking-bronco leaps, but with crow-hops or sudden jumps. I was never injured, just scared or bruised, and I soon dreaded each ride, afraid of her shenanigans. On one such ride, I landed in a sagebrush and cried—howled, actually—in front of all the men. The next morning, I locked myself in my room while the men saddled horses, and when Dad came to the house to get me, I called out that I was sick and couldn't help that day. I didn't want to tell him the truth—that I was afraid I would fall off again.

Dad just said, "Mary. You can ride that horse. She's not an outlaw; she's only playing. And we need the help. Cows can count, you know. You're coming with us." Dad kept me by his side all day. He explained how to anticipate the action when Trixie started to prance and shake her head. "She's just testing you," he'd say. "Tug the reins, now. Sharp, and quick. And then keep going; push her into a lope. Don't try to steal a ride. Let her know you're in charge, not just a passenger. Don't jerk her around, just ride her."

Trixie didn't make one false move until we started toward home at the end of the day and the men put their horses into a trot. Trixie kicked up her heels, and I got ready to hit the dirt, but Dad and the men started yelling like we were in a real rodeo: "Whoo-ee! Hang on! Pull 'er up! Ride 'er, ride 'er! Lean back!" I remember that magic feeling, the power of that instant I decided to *stay on!* "I can *do* it," I thought. I kicked her with my heels—a little scolding, really—and she jumped another step or two and then settled into an easy lope with the other horses. The men cheered and laughed, and I don't know who was prouder, Dad or me. I never had any trouble after that, and I rode Trixie for years.

Ranch kids didn't go to town much during the summer, but one Fourth of July Dad allowed Betty, Nancy, and me to enter our horses in the local Chuckwagon Days rodeo and parade. At home, we practiced and practiced for the barrel races, and on the big day Dad loaded our horses in the stock truck and away we went to the Big Piney Rodeo Grounds. As we unloaded, I spotted the other girls and their snorty, hot-blooded horses. Little brown Trixie seemed plain and dull, and I wished she would arch her neck or prance instead of just walking quietly in the parade and into the arena. I wished for a fancy shirt, too, and maybe a new hat or pretty boots.

When the barrel racing started, the fancy horses were all but runaways. They bolted out of the starting gate, taking big awkward turns around the barrels and then thundering toward the finish line. One of the girls nearly fell off when her horse veered at the flag, and another girl sobbed in fright when she couldn't control her horse.

When it was my turn, the starting flag dropped and Trixie ran smoothly into the pattern, hugging the three barrels like a pro, then racing for the finish line. I stuck on tight, thrilled to hear the announcer on the crackling loudspeaker say, "Watch that little horse, folks! There's what we like!" The crowd whistled and cheered, seeing Trixie for what she was—a tough, well-broke little ranch horse with lots of heart.

Proud as I was, that was the beginning and the end of my rodeo career. "I can't spend the summer hauling girls and horses. I'm glad you had fun, but I've got a ranch to run," Dad said. "And you'll learn a lot more on the ranch than you will in town, hanging around rodeos."

Actually, every day horseback was like a lesson of its own, far beyond the clinics and training seminars of today. Dad taught us to "read cattle," as stockmen say. Watch and listen. Look around.

Pay attention. "When you take a circle, ride up high from time to time so you can see everybody," he said, "and let everybody see you. That way we can work together."

"Don't bring a cow without her calf. Give her a chance. She'll tell you where he is." Sure enough, when Dad whistled, a cow picked up her head to look toward her baby, sleeping in the brush. "Sometimes," Dad said, "you'll have to look harder, because maybe she went to water, left him someplace with a babysitter cow, just like people do." He showed us how to spot a sick calf by its droopy ears. He told us to watch for calves who couldn't find their mothers. I asked him how a cow knew which calf was hers, since they all looked alike to me. Dad laughed, saying, "Don't you think your mother would know you in a train station?"

On one roundup, Dad told Betty, Nancy, and me to hold a bunch of cattle at the "Goat Ranch," a waterhole known by that name after an early homesteader tried to raise some goats there. Dad and the others rode back for more cattle. "Just wait here," Dad said, "and keep the cattle together."

We waited in the hot sun. Bored and impatient, we got off our horses to play in the dirt. Cattle grazed beyond us, but we gave little thought to those we saw wandering lazily across a little hill. Suddenly Betty jumped up saying, "Oh, my gosh! Those are *our* cows. See that one in front, the brockle-face one? We need to get them back before Dad finds out!" We scrambled to get on our horses. Too late, though—Dad and the other men were already riding toward us at a high lope, turning the cattle back our way.

When Dad reached us, he was furious. "I ought to paddle all three of you," he said coldly. "I ought to make you unsaddle and pack your saddles clear to home. If you can't do any better than this you don't deserve a horse." We girls barely kept from crying. We finished the day's work, and nothing more was ever said. Even now, though, when I ride with a lazy, disinterested cowboy, I remember Dad's words: "You don't deserve a horse."

Many cowboys and neighbors said among themselves that Dad was overly particular about cattle work and too fussy about details; a rebuke or correction from him humiliated adult or child. No one questioned, disagreed, or argued with his orders. Dad didn't shout, and he never allowed crude language. The other men followed that example—perhaps a concession to having to work with little cowgirls, or perhaps the manners of the times.

When Dad assigned positions on any trail drive, we kids hoped to ride with Bill, his friend and cousin. Bill was jolly and kind, and inside the pocket of his "jumper"—as denim jackets were called— he often carried sourdough pancakes rolled with jelly, which he shared with us. More important than the pancakes, though, was Bill's high-prestige job of "point man."

The point refers to the front of the herd, the lead, where the rider directs the movement of an entire herd of cattle. "If there's trouble at the drag, it probably started in the lead," the saying goes—meaning that the point man didn't pay attention or didn't know what he was doing. The job is never given to the unskilled, although many riders would like to be away from the dust and boredom at the drag, which is the back of the herd.

I jumped at the chance the first time Dad asked, "Do you think you can handle the lead?" I was thirteen or fourteen, perhaps, and I had ridden many a mile with Bill coaching me for just such a day. I could hardly wait to try! As the point man, I would decide where to apply pressure to speed or slow the herd, to steer them away from brush, to nudge the cattle without halting the drag. I would need to know if the cows were traveling "mothered up" with their babies. I would try to maintain a steadily moving mass, a string of cattle pushing and pulling each other by their own momentum. And it would be my judgment call to simply stop the herd if things went completely wrong. As usual, I was riding Trixie, and we would do our best.

Dad counted the cattle out the gate and cowboys fell into place, letting the cattle spill onto a vague, gently wandering trail that crossed shallow canyons and passed onto sagebrush flats. We would reach our destination in a few hours, we hoped. "If things are going right," Dad said, "let them keep traveling." He cautioned me about the gulch we'd pass beside and gave me a couple of tips: "The old cows in the front know where they are going, but the young ones don't, so don't let the lead get too far ahead. And the young cows don't keep track of their calves very well, so keep an eye on the drag. We can't have a run-back."

"Run-back" refers to a wild chase—complete pandemonium—when cows on the trail have separated from their calves. A calf's instinct is to go toward where he last nursed, perhaps miles back or hours ago. If just one calf separates from the herd and trots a few steps in the wrong direction, an alert cowboy can ease him back into the herd. A wrong move, though, can scare the calf into a run, and he'll be quickly joined by others and followed by a riot of bawling mother cows. It isn't a stampede—the cattle will stop and reconsider eventually—but hours may pass before the cattle are again under control and moving in the right direction.

I definitely didn't want to cause a run-back, or let the cattle get off the trail, or allow any other disaster. "One more thing," Dad said, just as Trixie and I were ready to go. He waved toward bulls pawing dust, bellering and snorting at each other, shaking their heads and looking for trouble. "Stay away from them," he warned. "Give them plenty of room. When they fight they won't even see you. A bull can knock a horse and rider flat."

I nodded, took a deep breath, and tried to look confident and grown up as I rode toward my position. I kept the cattle moving at an even pace, squeezing them here or there like toothpaste from a tube, pushing them beyond me, pressing a little crowd when it stopped, riding ahead of a few cattle that reached out too far. I looked over my shoulder constantly, watching for cues from the riders behind me. "Look back, look back," Bill would have said. "If you're doing it right, the cows won't know you're there. Be invisible. Pay attention, every minute."

My first day as point man was uneventful. I was relieved, thrilled really, when Dad said, "Good job, Mary. Didn't look like you needed any help. Some people have a knack. Glad you didn't stop them at the corner; things went just right." As years passed, I regularly took a place at the point, and many of those times held challenging situations. Experience taught me that the point man must pay attention every minute, even if it looks like he's just riding along, whistling. Keeping the cattle on the trail, or avoiding a run-back, isn't as easy as it looks. No matter, I have always savored the memory of Dad's praise for me that day.

"Just right" was Dad's highest compliment for a job well done. Those words ring back to me when cattle handle easily, when cowboys large or small are in the right place at the right time, when a day's work is achieved with artistry and skill.

Decades have passed since I was a little bitty cowgirl. I'm a grandmother now, and I've spent a lifetime teaching little cowboys how to do things right.

Recollections come a generation at a time: our son Tim—a little boy then—riding toward us, shouting excitedly, and waving his arm. "I killed a rattlesnake!" he shouted. We laughed and waited for him to catch up. "I couldn't get the rattles off so I brought the whole snake," he said breathlessly. "My horse didn't want to let me on, though."

Years later, Tim's little daughter, Anna, said to me, "Grandma, will you teach me to get on my horse by myself? Dad just heaves me up there, but he won't always be around." "Sure," I said, and when I told her about Tim and the snake, she giggled, saying, "This should be easy. All I have is a coat."

Last year, I helped my sister Nancy and her family with a big roundup, and we herded the grandchildren along with the cattle, reminiscing as we went. "Ride up high sometimes so we know where you are," we told the small ones, and, "Keep up, now. We don't want to lose you."

Nancy and I didn't scold the little bitty cowboys for their mistakes, as Dad might have done. We helped them on and off their horses, we answered their questions, we took pictures. The tired kids were revived by a drink of cold water, and when they began to play and tease, we told them, "Take care of your horses. They did all the work!"

We unsaddled at the end of the day, nodding toward each other, content to know the family-way ranching continued—with little kid cowboys filling big boots, riding together or riding alone, learning to read cows and do a day's work and feel the satisfaction of a job well done.

"Just right," I said. "Just right."

4

Love Story

"I'm lost," I thought. On a high stretch of sagebrush plains, the cattle ahead of me stopped to graze while I tried to make out a faint trail for them to follow. Alone except for the cows and my horse, I had no idea where to go. The time was the 1960s, and I was little more than a bride, having arrived here in the Big Horn Basin from ranchland in a different part of Wyoming. I had not explored much of this new country on my own and, at this moment, I was hot, tired, and annoyed at my plight. I couldn't see anything familiar—no landmarks I recognized, no road or trail in sight— not even a fence line. That morning Stan had asked if I would push a little straggler bunch of cows up the mesa trail to summer range while he tended to some other work. He would meet me "at the top," he said, but there was no sign of him now.

I turned my horse, took a deep breath, and squinted at a sheep wagon parked on a distant ridge, smoke drifting from its stovepipe.

I could see a saddled horse nearby—hobbled, I supposed—and I rode cautiously toward the camp. "Maybe I can ask directions," I thought to myself. I wondered who I'd find there—a sheepherder, I guessed, but I knew that many herders of that day were loners, drinkers, or vagrants, and some didn't speak English, so I was a little nervous as I approached. I looked around one more time hoping to see Stan, but when I didn't, I took another deep breath and headed toward the wagon.

To my surprise, a man and a woman stepped out of the wagon, hailing me from the doorstep. "C'mon over," the man called, waving his arm. "My name's John Ashton. This is my wife, Mary. We've got coffee." They didn't look scary at all. Relieved, I headed my horse toward them.

Before we had a chance for better introductions, a ragged pack of sheepdogs dashed out from beneath the wagon, snarling and yapping and tumbling over the ground toward me. John swore and yelled, "C'mon back here, damn you dogs!" But the ruckus startled the palomino colt I was riding; he spooked and crow-hopped, and we crashed through the camp, scattering buckets and firewood and tin cans everywhere. The horse had not really bucked, only jumped around, and I stayed on him easily enough, but it made a story for John to tell and retell for all the years I knew him: "That palomeeta horse bucked like Steamboat and she rode him to a standstill." That day, John laughingly picked up the mess and tied my horse near his at a makeshift hitch rack. He and Mary invited me again into the tidy sheep wagon, where Mary poured coffee into a thick white mug.

She insisted I take cookies with my coffee, scolding the dogs who by now were wagging their tails and begging for crumbs. She and John both talked at once, obviously enjoying the distraction of a stranger in their camp. I couldn't guess their ages except to think "not young." Mary was dignified and pretty, even if "a little heavy" as she often said when I knew her better, and she had lovely dark eyes, rosy cheeks, and curly black hair. John's round face beamed a broad toothless smile—I later learned that he saved his store-bought teeth for special occasions—and his blue eyes twinkled behind thick scratched-up eyeglasses. He was short and stocky, and on this day needed a shave, and when he took his hat off, I saw he was bald except for a rim of wild white hair above his ears.

They asked where I was going and why I was out there horseback, alone. I explained who I was and told them Stan had asked me to help out that day, thinking I'd enjoy a ride. "I was glad for a chance to be out horseback," I said, but then admitted, "I don't know where I'm going, though. It doesn't look like what Stan described."

"You're not far off. I can get you headed there," John said. "I'm working for Mercers now, but I know my way around these ranches; I've worked on most of them." He offered to ride with me awhile, and I gladly accepted. After we finished our coffee, I got back on my horse and waved to Mary, telling her I'd see her again someday. As we rode, John never stopped talking, and he showed me the long wand-like stick he carried, explaining how the bent nail on the end of it turned over rocks that might be arrowheads. "See? I don't even have to get off my horse!" He waved the stick toward the pasture boundaries and told me who the owners were. He pointed out the drainages and mountains that eventually became familiar to me—the mesa, first—the long plateau above the floor of Shell Valley, up and up toward Spanish Point, Black Butte, Trapper Canyon, Cloud Peak, Death Ridge, Battle Park, and other landmarks. (I had already learned the expression "the mountain," which is unique to this area; at first, I found it odd that natives would speak as though there was only one "mountain" on earth. Eventually, the phrase seemed normal to me, though, to describe the Big Horn Mountains as they rise dramatically from the terrain, making a chipped rim around a large bowl, the Big Horn Basin. The mountains themselves are nearly connected, wandering to become one long, contiguous lip above three counties: Johnson, Washakie, and Big Horn.)

Most herders barely owned their clothes, and they worked paycheck to paycheck. But I learned that John and Mary kept a little house and a permanent mailbox in the little nearby town of Tensleep. Mary accompanied John from job to job, often returning to Tensleep in the wintertime.

"You take care, now," John called out when we finally parted. "See you again, I hope."

I didn't see John and Mary again until several years later when he came to work for us. They moved to a house on our ranch, although by then they also had a tiny camp trailer of their own that their old International truck pulled around the range—a home on wheels, which was more comfortable than a sheep wagon. They stayed at our place for ten years or so, and John helped cowboy, tended sheep, did chores, fixed fence, and fed hay with us in the wintertime.

As Mary and I became friends, she told me about her own life. She'd never left the Big Horn Basin. Her first husband had been shot dead in a bizarre barroom incident that left her alone to raise their son. She never mentioned where she met John. John described his own colorful life as miner, cowboy, and handyman in Montana before he landed in northern Wyoming. Mary's son lived in our area, and I frequently ran into him, his wife, and their teenage son when they visited, bringing groceries, newspapers, and pictures. Mary showed me clippings of her grandson's achievements, and we shared the recipes she found in the papers.

At the ranch, John rigged up a little television for Mary. "Momma likes her programs," he would say. One of Mary's favorite shows was a sin-and-salvation evangelical program, featuring the "PTL Club, Praise the Lord!" Stan and John teased her about her fondness for the zealous preacher and his constant pleas for money. "Momma, are you sending *our good money* to that shyster? The Lord helps them that helps themselves!" Mary scolded Stan and John about their own slim chances of reaching heaven, but she good-naturedly refilled their coffee cups.

Mary wasn't interested in livestock work or the out-of-doors, but she liked to spend the summers with John "out t' camp." She passed the time in the trailer listening to a little battery radio. She wore an apron over her pretty blouse and a scarf tied around her hair curlers. She kept her nails manicured and her makeup freshly applied, and our children marveled at her meticulously penciled eyebrows. John fussed to keep her comfortable, making sure the campsite stayed neat and clean. After each day of work he hurried back to her—she'd be waiting, perusing a stack of old newspapers. "By the time I get my papers read," she'd say, "I just don't get a thing done."

On the range, John herded the sheep conscientiously, keeping them on good grass so we'd have fat lambs for market. One summer morning he showed up before daylight at our house, white-faced and breathless, shaking, close to tears. He'd driven twenty-five miles from the mountain camp, Mary beside him. He said, "Boss, you've gotta come. Something's been in the sheep, maybe a bear! I heard them in the night. There's dead ones everywhere. God, it's awful." Stan followed him back to the range, where they saw how the sheep had spooked from the bed-ground, stampeded off a knoll, and tumbled over each other into a hideous pileup. More than a hundred died on the spot—suffocated or turned on their backs. Stan shot others as they thrashed and blatted in a sickening heap.

We called the government trapper and the game warden; they trapped a black bear the next night, verifying John's suspicion. Stan used a bulldozer to bury the rotting, smelly sheep carcasses. Financially, the loss was huge. Emotionally and personally, it was as hard for John as it was for us. "Nothing you could have done," Stan told John. "It wasn't your fault."

In a different year, during a record-breaking rainy spring, we suffered a different loss. Mary sat primly beside John in the old truck as it roared in from the south range. "Boss, I need help. I don't know what to do. There's cows dyin', just layin' there dyin.' I don't know why." We followed John back out to the range where we discovered thirteen dead cows—all with nursing calves that we had to rope and haul home to raise as orphans. We learned that the lush salt sage had blossomed uncharacteristically when the cool, wet weather suddenly turned hot, and the cows died from bloat. We moved the rest of the cattle into a different pasture and have never known that to happen again.

John and Mary shared our misfortunes as if they were their own, but they encouraged us with smiles and laughs, too. "Here, you need your vitamins," John often said, shaking a cigarette out of the pack. "Let's have a cigareet," he'd say, when the cattle slowed down on the trail. Sure enough, by the time the "cigareet" was finished, the cattle were ready to move on. Anytime we passed their camp, Mary called out, "May I serve you something?" In grimy Levi's and boots, Stan and I squeezed into the trailer's dinette seat for coffee and cookies, where Mary primly set out napkins and teaspoons for us, just as though we were royalty.

The game warden stopped at the mountain camp one time, just passing by. "Now get the hell out of here," John told him. "You've got no business here. You're trespassing. Stan's the boss of this outfit. If Stan ain't here, his wife's the boss, and if she ain't here, I'm the boss, and I'm tellin' you to get the hell off this place." The warden was a decent fellow who was just trying to do his job, so Stan called him, hoping to smooth things over. The warden just laughed, saying, "I tried the best I could to calm that old man down. Was he excited! I never took such a chewing-out! He had nothing to hide as far as I could tell, but he was sure looking after the ranch." Whenever Mary spoke of the incident, she defended John's heroism, for his standing up against the "red-shirt lawman that ought to mind his own business."

Occasionally Mary and John chugged to town in their old green truck to do their laundry and pick up the mail or a few special groceries. Those trips were risky if they stopped for "just one drink," as one drink often led to another and another. During the County Fair one year we saw the truck parked crookedly on Main Street in front of the Rustler Bar, and we knew they must be inside swapping news and stories with friends. We had 4-H kids with us so we didn't stop by the bar until after the steer show, several hours later. By then John and Mary were dancing tipsily to the cowboy band. They hailed us at the door and bought a round of drinks, and when we left at midnight we saw them rolling dice, laughing, still partying. They always made it back "out to camp" safely, but sometimes those barroom stops led to quarrels. One early morning when Stan and I drove into their camp, John lay sprawled face down outside the trailer, snoring in the dewy grass. He had drunk too much in town the night before and had fallen, trying to walk from the truck into the trailer. "Serves him right," Mary said. "I didn't look back." She made coffee, and John rallied to join us, and they quickly forgot all about their quarrel.

Mary was ill most of one spring, and John took her to a doctor who then sent her to the hospital for tests. A few days later John drove home from town, right up to the corral fence where we were working. He climbed out of the old truck, left the door ajar, and stumbled over to lean on the fence. "It's Momma," he said. "It's The Cancer. The Doctor said so. It's bad, it's bad." And he put his head down, put his face in his hands, and began to cry.

Mary died without ever coming home from the hospital. John tried to stay and work, but without her presence as his compass, he could not hold his life together. He began to complain about the work and look for reasons to go to town, where he started to drink more. When John, Stan, and I worked together on the mountain one day, Stan spotted some cattle on a ridge where John had ridden—cattle John should have gathered. "By God, I didn't miss any cattle," John said angrily. "John," Stan said as he patiently pointed toward the hill. "Look. You can see them from here. It's no big deal. Just go get them." John squared off to ask Stan, "Are you calling me a liar? That's all, then. Nobody calls me a liar. I quit." John rode his horse to the corral, turned it loose, threw his saddle into the old truck, and drove away.

We couldn't change his mind, although we tried. Somebody helped him move his things out of the house, and he disappeared.

He came back to ask for a job the following spring, but by then we'd hired a local woman and a Mexican fellow for calving help. John was hurt and embarrassed when we couldn't use him, and after he left, he told people, "Women and wetbacks, that's all they hire out there now. Women and wetbacks. I wouldn't work there."

Years have passed. When I ride past familiar places on the range where John and Mary might have parked their trailer on a certain hill or beside a stream, I smile, wishing for that cup of coffee and a "cigareet" and news from the week-old paper. I remember John and Mary's sweet loyalty to each other and to us. For an instant I can almost see John's toothless smile, and I pretend to hear his proud announcement, "Come on in, Momma's got some pie!"

5

Sheep Country

In 1910, roughly 5.5 million sheep were said to be using Wyoming rangelands, grazing on government lands in addition to homesteads and private holdings. Those numbers diminished quickly as the country became overgrazed or "sheeped out," as the common phrase implied. The Taylor Grazing Act of 1934 restructured the use of federally owned lands, limiting livestock numbers and declaring responsibilities for the ranchers who wished to participate in the plan, whether with sheep or cattle. (Prior to that time, competition for range grass was fierce, and great animosity prevailed among cattlemen and sheepmen. Those conflicts are well described in range-war tales and documentaries. Maps of western states show numerous areas called "Deadline Ridge," or "Deadline," which denote the line that sheep must not cross. Those are stories for others to tell as, in my own life, I never witnessed those angry scenes.) Now, a century later, there are fewer than 500,000 sheep in Wyoming.

———

Stan's father, Howard, bought sheep in the 1930s and brought them to the ranch in Shell Valley. He had determined that running sheep together with the cattle he owned would balance his income column between market spikes and dips—when cattle were up, sheep might be down, and vice versa. Sheep offered two yields each year, too: mutton and wool. Howard was one of many ranchers who adopted this strategy, which worked successfully for many years. The Flitner ranch supported two to three "bands" of sheep; generally, a thousand sheep comprise a band. Each band called for a herder and an accompanying sheep wagon, as most of the range was not fenced, or "sheep-tight." A herder made sure the sheep stayed together and found good feed and water, moving to new locations every few days.

Here in Shell Valley, ranchers "shed-lambed," which means that lambs are birthed inside a giant shed, under the supervision of a lambing crew. In other places the norm is to "lamb out," letting the ewes lamb unattended on the range, and letting the strong survive. There are pros and cons to these management options: "lambing out" is not as labor-intensive, but the lamb losses will be higher.

In a shed-lambing, when a ewe hasn't sufficient milk to raise a set of twins, her extra lamb can be grafted onto a different ewe whose weak lamb has died. If the extras don't all find mothers, the leftover lambs are often bottle-fed and raised by the ranch kids, and in my experience and by most accounts, these "bum" lambs are the cutest *and* the most aggravating of all the Lord's creatures.

One year when our kids were small, they were put in charge of six or seven of these bums, which they named, fed, tended, and kept alive, contrary to a sheep's seemingly born determination to die. By late summer, the lambs weighed about eighty pounds each and followed the kids like house pets, all over the ranch. Those bums were absolutely everywhere—in the flowerbeds, the yard, the garage, and even the house if someone left the door open— usually *out* of their pen and *in* the way. They received a cussing several times a day, accompanied by my threats to turn them into a big lamb stew.

The rest of the sheep—ewes with lambs, and yearlings—went back to the range or to the high country to summer. Sheep payday came in August, when we shipped the market lambs straight off the mountain range. The herder trailed the sheep to our corrals, and on the following morning, we'd sort a thousand or so lambs away from the ewes, then trail them down several miles where they could be loaded onto semitrucks. It would be a busy, trying day, because mistakes could result in extra "shrink," or loss of pounds on the lambs, and thus fewer dollars.

We always expected trouble at the Rockpile—a half mile of boulders nearly vertical, with a trail winding down through the treacherous ledges. This year, after all was said and done, nobody would admit suggesting the plan of action: to use the bum lambs as Judas goats or lead sheep, as in the expression "follow like sheep." In the plan, our three older children, Carol, Tim, and Sara, would lead their favorite pets, Toby, Jumper, and Snuffy, and the other half-dozen bums would follow, and then the other thousand would, too. It seemed like such a great idea to haul the bums up from home, and the kids were excited about being a part of things.

We slept at the log cabin on the mountain the night before shipping day, and we started work at dawn on that cold mountain morning—corralling the sheep and then separating the lambs from the ewes before we drove the lambs toward the trail. All seemed perfect. I remember what a beautiful sight it was when the sheep moved across the hillside—a white, sprawling mass, spilling over Snowshoe Pass. Our littlest boy, Dan, was in the pickup truck with me, and my husband, Stan, and the other three kids walked confidently with the herder and his dogs. Practically a scene from *Heidi*.

At the Rockpile, the plan fell apart immediately: the band of sheep looked at the rocks and began to mill in noisy, bleating pandemonium. Carol, Tim, and Sara couldn't drag their pet lambs off the first ledge—Toby and Jumper and Snuffy pulled away and escaped, disappearing into the enormous, jostling herd. Stan and the herder yelled and swore, the dogs barked, the kids cried—and I watched helplessly from the pickup at the top of the pass. At the height of the confusion, Stan somehow grabbed one of the other bums and wrestled it the first few steps down through the rocks; a few sheep did follow, and then a few more, and, finally, the rest of them clambered down the trail.

The semitrucks were waiting at pens near the Shell Ranger Station, a mile or so farther on. While the men wasted no time loading the sheep, the kids ran from pen to pen calling for their bums, peering into all the corners, and standing on the fence to see if they could spot them.

Among a thousand look-alikes, the bums weren't to be found. The chute gates banged shut and the trucks pulled away, leaving the kids sobbing in the empty corral.

To make matters worse, the sheepherder, Joe Burks, had found a bottle of whiskey he'd stashed for the occasion, and he was swigging down Southern Comfort and mourning his loss. "Poor little lambs," he said. "I took good care of them. Yup, by this time tomorrow they'll be hangin' on a meat hook." That wasn't even true, as the lambs were going first to a feedlot, but there was no convincing the children and they cried even harder. I tried to tell them that their grandpa and dad would find the bums in Worland at the scales, where the buyer would weigh the sheep and write the paycheck, but we were a gloomy crew heading off the mountain in the pickup.

When Stan got home, he had to tell us he'd failed to find the bums. The kids were heartbroken, and Stan and I were disappointed, too. "I tried," he said. "I really tried. I looked through every damned pen. I just couldn't find them. I can't figure it out."

We sat down for supper, but nobody was hungry, and the kids kept glaring at us. When the phone rang, Stan jumped for it, anxious to escape. We heard him say, "You did? Really? Well, I'll be damned. We'll be right up. Thanks."

He shared the news—Toby and Jumper and the other bums were fine, grazing around the Shell Ranger Station lawn. In short order, they'd had enough of being real sheep; they had separated themselves out from the bunch before we ever got to the corrals, and they found a spot to rest. The district ranger found them on the porch at the Ranger Station that evening, and he was anxious to get rid of them.

Joe went back to his wagon with the rest of his Southern Comfort, and he would stay on the range with the sheep until they came back to the ranch for lambing. Joe was a good herder, and a short "bender" was deemed acceptable for sheepherders. For one thing, herders were hard to find. Most of them were loners, misfits, has-beens, drunks, or eccentrics of some type or another—not all, but most. Most ranchers didn't ask questions, if the guy could herd sheep at all. The job didn't require too much—living out on the range in a sheep wagon, herding the sheep to a bed-ground at night, and putting them out to fresh grass each morning.

We had hired Joe sight unseen. One herder had quit us without warning, and we'd left word at all the bars in the area to let us know if someone showed up looking for a sheepherding job. The Mint Bar in Sheridan phoned to tell us a guy named Joe Burks was there, wanting to go to work. Actually, Stan and I were hosting a party at our house that night, and when the phone rang, Stan called out, "Tell 'em I'm not home!" When I said, "It's the Mint. They say they've got a herder," Stan said hurriedly, "I'll take the call!" We sent a bus ticket—a fairly common procedure at that time—and Stan met the man at the station in Greybull.

Joe arrived as we expected: unshaven, dirty, smelling of bar smoke and whiskey. Sick, sober, and sorry, without a dime to his name. He wore cowboy boots and a black, sweat-stained cowboy hat that was as dirty as the rest of him. Oddly, his gear included a scarred-up Connolly saddle, a set of gal leg spurs, some short, fringed chaps—"chinks," as they are called—and a coiled lariat rope with a *square knot* holding it together. "That was a surprise," Stan said. "I'd have expected a guy to pawn that stuff to pay the bar tab. And the knot? Never saw such a thing."

Stan hauled him out to the camp, showed him the horse to use, fed the sheepdog, gave Joe a brief explanation of the range, and left him at the wagon, hoping to God the guy would sober up before the sheep strayed completely away or, worst of all things, mixed with the neighbors' sheep.

The kids and I often went along with Stan to "tend camp," which involved bringing supplies to the herders. Usually, I had a child on my lap, and the others bounced along in the back of the pickup. If it was a day that included moving camp, Stan would load the firewood that hadn't been used up, the horse hay if there was any left, the buckets, the doorstep, and anything else that was lying loose, then hook the wagon onto the pickup and pull to a next location where the herder would meet us with the sheep. We were with Stan the first time he drove out to check on Joe.

When we arrived at camp, the stove was cold; the dog, the horse, and Joe were gone. We wondered aloud if he'd simply quit and then ridden toward town, but within a few minutes, we saw him riding along the ridge toward us. The bald-faced camp horse was stepping smartly, moving with uncharacteristic style and purpose, and Joe was riding tall and straight, wearing his chinks and the spurs. He'd put the sheep out early, he said, and he was just coming back to the wagon to make some coffee. He still looked sick and shaky, but he'd cleaned up somewhat. He tipped his hat to me and said, "Ma'am?" politely, and smiled at the children. As they got to know him, the kids learned to love Joe, in spite of his crooked, broken face that someone said looked like "the map of the Big Horns."

Usually a cowboy would view a sheepherding job as the rock bottom of life, a humiliation of sorts. We wondered how Joe came to be herding sheep, and we asked around, curious about his circumstances, which didn't seem in keeping with his classic saddle and cowboy gear. We learned that he was well known on "the other side of the mountain," the ranch country near the Montana state

line. We heard from people who knew him in Billings, Hardin, and
Sheridan, who told us, "Burks? Hell, yes. Now *there* was a cowboy.
Best hand I ever saw with a horse." "Herding sheep, you say? I'll be
damned." "Wondered what ever happened to him. Last I knew he
was tangled up with a dude gal, and then the bottle got him, I guess."

He had married the "dude gal," as I learned when he showed me
a photograph of several young cowboy dandies, well dressed and
laughing, standing together. A pretty, dark-haired woman held his
arm, smiling up at him. When I asked who she was, he told me
she was his wife; he took a big pull on his cigarette, blew out the
smoke, and said, "Just didn't work out."

His life went on a downhill slide after that, people told us, and
his drinking got in the way of the good honest hand he had been.
He was apt to take off in the middle of a roundup, headed for
town. Once he was said to have turned loose a herd of horses on
the range—just left them—while he'd gone off on a drinking spree
with a buddy. After that nobody wanted to hire him, and that's
when he started herding sheep.

His past didn't matter much to us, as long as the lambs came
in fat and the range remained unscarred for its seasonal rest. Joe
worked for us for ten years or so before he left the way most herders did—finding an excuse to quit and go to town. We heard that
he spent (drank up, gave away, or lost) $1,700 in four days.

After that, we only saw him a few times through the years, but
I think of him often. Nowadays, horsemanship clinics are frequently offered by professional "horse trainers." Joe would have
scoffed at such a word as "clinic," but he had that gift with horses:
he could make any old plug of a horse travel like a champion,
steady and alert, ears pricked. Colts responded to his quiet ways,
and although he never got on a bronc in all his days at our place,
he willingly shared his knowledge. His philosophy still works:
"never spoil them, never let them get away with anything, and
never steal a ride."

Tim, a skinny little kid then, worked his first bronc—a two-year-old colt—in the round corral at the ranch while Joe propped his gimpy leg on a rail, observing the colt's behavior and offering Tim encouragement. Joe's gnarled hands shook as he rolled a cigarette, but his voice was steady, soothing the young horse and rider. A gap-toothed smile lit up the wrinkly old face, beaming with enjoyment on a fine spring morning—watching that wonderful combination of boy and horse.

As we worked our way through the years of sheep ranching, other herders came and went: Joe Martinez, who made the best beans and tortillas and kept his wagon spotless; Andy, who bought two new pairs of Levi's in the fall, wore one over top of the other until spring, and then took the top pair off and threw them away; Eddie Velasquez, who wore a black-and-white pinto horsehide vest. These men rotated among the ranches of our area, not always staying put, but usually showing up on neighboring places—in and out, back and forth. The children were sometimes fascinated with these strange characters—and they liked most of them—but they saw little of them other than the excursions to sheep camp with Howard or Stan.

Many of my old journal entries record the fun the kids had when they went along with the camp tender. Sometimes Stan or Howard would let them practice driving the pickup, or sometimes they could shoot the .22, or they might snack on some of the store-bought cookies—gingersnaps, usually—which the ranch kept in stock for the herder. They could play with the sheepdog, or dig worms, or as happened once, catch a mouse in the woodpile and sneak it home in a pocket! A couple of times when the truck quit or got stuck, the whole crew had to walk until they found help. The stories had happy endings and lots of laughs in the retelling.

In my journal I titled one of those old stories "The Best Day." On that day, Stan asked if I'd like to take a horse and gather up a little bunch of ewes that had escaped from the band being herded to a new pasture. "It should be easy, and maybe Carol could go along, ride old Peanuts. It would save me some time," he said, "if you could push those ewes to the top of the mesa while I'm moving the sheep wagon and setting the herder's camp." I liked the idea of a day away from our houseful of kids, and Stan's mom said she'd watch the little ones. Carol, our oldest at five years old, was dying to go. Stan hauled the two of us and our horses to the bottom of the mesa and waved us toward the trail.

"You can't go wrong," Stan said. "There's one canyon that comes in; don't take it—just stay in the bottom till you hit a little spring and some old wooden troughs. Let the sheep water there and then let 'em climb out that next fork to the northeast. You'll end up at the Split Ear Corral and I'll meet you there."

We found the sheep, and they climbed agreeably ahead of us, up the steep, narrow Split Ear Canyon trail. We made slow progress; I helped Carol off or on her horse, or tightened her cinch, or reached for her sandwich, or tied her coat to the saddle. Crowding the sheep when they slowed, or pausing at the scary places, I coaxed Carol along, hoping I was right in saying, "Trust Peanuts. He knows where he's going."

I nervously watched rainclouds gathering and helped Carol zip her coat when a knife-cold wind swirled behind us. "We're almost there," I said, but I really didn't know. "Split Ear," Carol said. "What a funny name. How come they call it Split Ear?"

"I'm not sure," I said. "Someone told me the old-timers used this corral to brand and earmark their cattle, splitting their ears with a knife to identify them." I watched the clouds, trying to be patient while Carol chattered endlessly about Indian Paintbrush, pretty rocks, the dog, the horses, and every kind of trivia. I tried not to say, "Pay attention. Hurry up."

Finally, the narrow trail took us out of Split Ear, just as the sky turned black and lightning cracked ahead of roaring thunder. On top of the flat, sage-covered mesa, sure enough we spied the pole corrals, and a little farther on, the rattletrap blue two-horse trailer that Stan had left there as he promised. We dropped the sheep as Stan had instructed, and I cried out, "Come on, Carol. *Hurry. We're going to get wet!*" We got to the trailer's shelter just as rain and hail caught us.

The horses turned their rumps to the storm; I quickly tied them against the trailer and pulled Carol inside. The rain hammered the metal roof while the old trailer shook and rattled with the wind. Carol began to cry then, frightened by the fury of the storm, and I hugged my skinny, shivering little girl. "Will Dad find us?" she asked. I tried to sound confident: "Of course. He'll be here any minute." I wished for the hot coffee Stan might be drinking with the herder and wondered why he was so slow—what would I do if he had trouble, didn't show up? Miles from home, late in the day. No cell phones then—the 1960s—and no way to find help. Stan's mother would be worried. "Would anyone come looking for us?" I wondered. Perhaps after the rain we could ride home, although it would be a fifteen-mile ride. "No," I thought. "We'll wait." We weren't in any danger, really, just uncomfortable and annoyed. "Here. I found a candy bar in my pocket." Reassured, Carol kept talking, talking, talking. I managed, "Mmm-hm. Mmm-hm. Yeah, I think so," to her questions.

At almost dark, Stan showed up, the 4-wheel-drive truck grinding through the mud. "Roads got bad," he said. "I had to chain up, damned near got stuck. Herder wanted to gab, took too long. We need the rain, though." We rehooked the trailer, loaded the horses, and headed home, downhill now all the way. The heater roared full blast. Carol sighed and yawned, snuggled between us in the truck seat while the wet dogs huddled on the floor. "This was the best day of my life," she said.

I believe there were many "best days" for all our children and others, growing up on the ranch. At any rate, our family is among those who no longer raise sheep, but I respect that proud industry and its own place in ranch culture. In Shell Valley, only a couple of farm flocks remain. Not many bona fide sheep ranches still function in Wyoming; few have survived the economic struggles of down markets, increasing losses to predators, and the development of synthetic fabrics and products that have reduced the demand for wool. In addition, labor problems, fencing issues, government regulations, and the general intensity of the work have resulted in the decline of sheep numbers here in the Big Horn Mountains and the larger regions of the West.

The lambing sheds are dilapidated now; weeds grow up in old corrals. Sheepherders don't frequent local stores to cash their paychecks. Empty sheep wagons stand parked in a few backyards, unlike my memories of the years when wagons dotted the far ridges of the Big Horn Mountains—a viable, productive presence.

Sometimes we see a shearing set-up or we notice some shed lights at night, and occasionally we see a sheep camp still in use. Last summer, it was delightful to talk with a sheep rancher as he followed a band of sheep trailing through the Big Horn Mountains with his little kids beside him—just as we had done, generations ago. I have great admiration for the sheep ranchers who succeed. And I wish them good luck with the bum lambs.

6

Betting It All

"I don't know why we did this," Delmer said glumly, chewing nervously on a toothpick. He paced alongside the country racetrack. "I'm not a gambling man." A short, wiry man with leathery skin, he tugged the brim of his cowboy hat and scuffed the toe of his boot in the dirt.

In the gathering darkness, a small crowd of us stood together waiting for a horse race to start. We were fifty miles from home in the little Wyoming cowtown of Tensleep, where the traditional Labor Day celebration was in full swing. Things had not gone quite the way we had planned. At the end of the day, we'd made a reckless bet with the man we now waited for, and most of us were secretly hoping he wouldn't show up.

Earlier that week, in September 1978, Stan and I had declared a holiday for ourselves. We were tired and discouraged—behind in our summer's work and afraid we'd never be otherwise. The hay had been rained on, our fences were slack, and our livestock was scattered in pastures where they didn't belong. The bankers were leaning on us to provide income estimates and inventories—asking for meetings that we wanted to avoid because we knew there weren't going to be good answers. Livestock market prices were low, and our loan payments were coming due. Interest rates were high, too. Like many other ranchers, we didn't know how we were going to get through our next year. We felt dangerously close to complete financial failure; still, we agreed, a day away from the ranch wasn't going to affect the big, bleak picture we faced.

We called some friends to see if they'd go with us to the races, which were always part of Labor Day action in Tensleep. "I've worried as much as I can worry and it doesn't help," Stan said to our pals, most of whom were in the same boat themselves. "I've worn out the eraser on my pencil, trying to get numbers to add up right. I'm tired of it. Let's go have some fun. Maybe it will change our luck."

Our ranch neighbors Delmer and Marge said they'd come too, and they brought a couple of horses we knew to be extra fast. Their handsome cowboy son came along as jockey. With several other couples, we formed a jolly caravan driving along the Nowood River road, sharing laughs long overdue. Indeed, it was a "great day for the Race" our friend Tom said. "The *Human* Race," we shouted on cue. We couldn't fail to have fun in the shirt-sleeve weather—sunny and bright with vibrant red and gold foliage and a blue, blue sky.

People brought horses to Tensleep from both sides of the Big Horn Mountains, mostly ranch horses with some speed—no professionals—and the races offered modest entry fees and prize money. Spectators came to see friends, bet a little money, and watch good horses run.

We parked beside trucks with saddled horses tied to them and we wandered easily among people wearing Levi's and cowboy hats, drinking beer and eating barbecued beef sandwiches served up by the local women's club. No one drove shiny diesel pickup trucks or pulled fancy horse trailers, no one had big money to throw around, and everyone was congenial and friendly, ready for fun.

The racetrack was an old dirt landing strip outside of town. It didn't use a starting gate, just a line in the dirt. Each race began with the drop of a flag—then the quick quarter-mile run, four horses each time, and after several elimination heats, a final run-off for the winnings. We made some money when our horses won their races, and we won a few side bets as well. We laughed and joked about being the Shell Creek Syndicate, enjoying good-natured rivalries with folks from nearby towns—Basin, Buffalo, Worland, and Hyattville.

After the races, our little syndicate stopped to celebrate with the crowd at a Tensleep bar before heading away; we'd still be back at the ranch before dark, we thought. Our kids had stayed in Shell to play with cousins and neighbors, and we'd planned to be home early. We elbowed our way into the crowded bar, a smoke-filled, noisy, old pine-walled joint with a neon Coors Beer sign blinking above the bar and the jukebox belting out country songs. The barmaid hurried from table to table with her tray of drinks, taking cash and making change from the apron she wore. We found a place to sit and joined the fun.

Inside the bar, though, a big noisy fellow wandered annoyingly from table to table, bragging about a horse he had back at his ranch—faster than *any* of the winners, he said, and he wanted to bet $1,000 cash to prove it. He kept heckling everyone and somehow we heard ourselves saying, "Well, go home and get him. You're on. We'll take your bet." He left to get the horse, telling us to get our money ready—he'd be right back.

So now we waited at the track, and no wonder Delmer was nervous. We all were. Our talk had been too tough, we agreed; when we pooled our money to cover the bet, our syndicate only had $300 of the $1,000 cash we needed. We emptied pockets and purses and dug through the jockey box of the pickup, and that was all we could come up with, even with the money we'd won earlier in the day. Cashing an out-of-town check on a holiday was difficult, and ATMs had not been invented yet. The bartender reluctantly agreed to honor our friend Tom's check for the remaining $700, and we headed to the track with the money and our horse, each of us wondering if that other horse really could outrun ours and how to manage if we lost. With a lump in my throat I worried about buying groceries and paying school lunch money, which had been raised to $8 a month that year!

Suddenly, Stan turned to Delmer. "What the hell, Delmer. This isn't betting it *all*, like ranching is. You don't think we're gamblers? We've been gambling for years—gambling that the cow business is going to get strong again and that we're going to make it through. We're gambling that it will rain, or that cattle prices will come up, or that we'll heal up after we've been bucked off horses or run over by cows. We've walked home when the truck broke down, worked day jobs, made a living against the odds. We've gambled for high stakes, higher than this horse race, for sure."

"Too late now anyway," Tom called. "Here he comes. We're entered. And it's a great day for the Race." No one laughed. I stood with my hands clamped together to keep them from shaking, and we all watched as the man unloaded a big, leggy Thoroughbred from his truck, and we sized him up. "Oh man," Tom whispered. "He's built to run." Each of the two riders handed $1,000 cash to Dave, a local rancher who would flag the start and declare the winner. He rolled the bills up tight and snapped them with a hard "click" into the pocket of his orange plaid polyester shirt. As the horses pranced toward the starting line, people from the bar angled for a place to watch.

When the flag dropped, the horses jumped out neck-and-neck, hooves pounding the dirt and kicking up so much dust that we could not see the finish line. We didn't know whether we had won or lost until we heard Stan cheering at the far end of the track. "We did it! We won! We're okay!" We jumped up and down and clapped each other on the back to congratulate ourselves, but we felt more relief than triumph, and we gathered up our money and left town as quickly as we could. We knew that a $1,000 was a lot of money to the loser, too.

I didn't know at the time how prophetic that conversation was, about "betting it all" for our family's future, and "too late now, we're entered." For a while, there was no way to get out of ranching, even had we wanted to do that. Many ranches were on the market, but few were selling—ranchland had no recreational value then, no glamour or prestige for millionaires or absentee owners. Bankers described ranchers like ourselves as "land-rich, cash-poor," an expression that referred to the worth of a ranch on paper but that did not manifest into cash for paying debts.

I saw our starting-line judge Dave at a community picnic this past summer; we laughed and reminisced about that thrilling day, and I couldn't help but notice that his once-red hair and mustache were more gray now than red. "More than thirty years ago, just imagine," Dave said. "Lots of changes, since then." We spoke of neighbors who made it through the tough times and others who didn't—Delmer and Marge were among those who sold out and moved away. On Shell Creek and throughout the Big Horn Basin only a few of the ranches of that time remain intact. Many others are divided into residential parcels, without agricultural purpose.

Our conversation honored the sadness brought by the passing of old friends, and we shared personal disappointments from politics to business deals that didn't go our way. We laughed as we talked about the fancy modern tractors and ranch equipment people use nowadays, and the convenience of computers and cell phones. I recalled how proudly we'd rattled our old blue truck into Tensleep that September day, pulling an open-top horse trailer.

"I never forget," I told Dave, "that it could have gone either way for us. We could have lost, just as easily. Some good people went under, or chose to bail out when they got a chance." Dave nodded in agreement, and we sat quietly for a few moments, remembering how each of us had seen record high prices swing to record low more than once during the past decades, and we acknowledged that we'd likely see that swing again. "Since then," I said, "Stan and I have spent a lot of days trying to figure out how to be prepared when that happens next time."

"We worked hard," I said, "but that wouldn't have been enough—we had to ride it out. It took determination—and moisture, of course, and, at last, a favorable cattle market—things bigger than we could control. Our banker stuck with us."

"We did a lot of crazy things to make a little extra money and keep holding on," I added. I reminded Dave that we'd gotten an outfitter's license so we could take in paying customers who wanted to hunt elk or deer on our land. With a friend, Stan contracted to build fence for profit, selling electric fence supplies to sweeten the deal. When our girls reached high school, they waitressed in town in addition to doing ranch work, and the boys put in long, hard days working side-by-side with Stan and me. "We never gave up. We did every little thing we could to make a few dollars. It all added up."

Dave nodded, smiling in agreement. "Yes, and we *did* have some fun."

It was a gamble. Things could have gone the other way. We bet it all, and we won.

7

The Winter of 50 Below

When I hear stories about record cold, one winter comes to mind.

In the late fall of 1978 the weather turned bitter, with snow and more snow accompanying below-zero temperatures. Stan and I left the cattle on the range as long as we dared, but the weather kept getting worse, and the snow piled up and up until it became too deep for the cattle to find grass, even on south-facing slopes.

Late in November, two months ahead of plans, we gathered the cattle from what we'd hoped would be winter pasture, and we headed them toward home. It took several days to find them all, even with friends helping us, and our horses struggled to break trail, ten miles through crusted snow a foot deep.

"They've got to have feed," Stan said. "We haven't got any choice. We can't leave them out there." I knew we didn't have enough hay to feed for two extra months, and with such grim weather, each rancher was likely to hoard his own supply. Even if we could locate hay for sale, the prices would be sky high.

We heard of a farmer who had hay for sale in Basin, a small town on the Big Horn River, so we called him and bought what he had, dangerously pushing our credit line at the bank. "Hay's like money in the bank," ranchers say. "You need it when you need it." We didn't have enough of either, but we wouldn't let our cattle starve, and the cattle market was too low for us to think about selling them. We didn't have a dependable truck and thought it too costly to hire someone to bring the hay around to the ranch, so in early December we trailed the cattle down the highway and along a back road, fifteen miles over to the man's farm, where we could feed them from his haystack. We took our team of Belgian draft horses and hay sled over there to use, and our old friend and hired hand, John Ashton, agreed to stay there in his little camp trailer to keep an eye on things.

"Wish we had one of them big tractors," John said.

"At least I know these will start every day," Stan said in defense of his horses and the sled. Watching our neighbors fight against frozen engines, jumper cables, dead batteries, and mechanical problems all winter long proved Stan right.

I was worried sick when I saw where the cattle were going to be—I had heard awful stories about cattle falling through the ice on the Big Horn River. We had just gotten the livestock settled there when the weather went from bad to worse, and the river ice froze two feet thick. I need not have worried about anything falling through the ice. Snow kept falling, and the temperatures kept dropping throughout the Big Horn Basin. The recorded low that winter was 52 degrees below zero. The snow depth leveled at nearly three feet. Wind blew, too, and drifted the back road shut, closing our "short cut." We faced a fifty-mile round trip each day in the pickup, from home to the farm in Basin, just to get the cattle fed. My ranch journal tells the story:

> February 3, 1979:
>
> Thirty-two degrees below zero this morning. The weather has held like this for over thirty days now and no end in sight, the radio weatherman says. He calls this a temperature inversion, an enormous puffball of frigid air that is stalled here in the Big Horn Basin, and he says it will remain until a big wind blows it through the layer of warm air above it. I am not interested in the details.

We are in a God-awful routine. Each morning we start
the pickup, which is parked in a shed and plugged into
a head-bolt heater overnight. We load the four kids into
the cab of the pickup, along with the two dogs—heater
going full blast—and we meet the school bus at the end of
the lane, a half-mile from home. The kids hop on the bus
and we continue on, crossing the river bridge at Greybull,
then eight more miles to Basin to reach the cattle. The
radio blares out the market report (bad), the news (bad),
and the weather forecast (bad). I'm tired of hearing it.

We pull in to the camp where old John, bald and
toothless, waves through a peek-hole in the trailer's frosty
window, signaling that he has hot coffee ready. His dog
Cactus crawls stiffly out from under the step, barely
wagging his tail, too cold to show enthusiasm. We shake
out of our coats while John pours the coffee. "Coffee with
a bead on it," he likes to say. "Have a cigareet. You need
your vitamins." He has on his insulated coveralls; I think
he doesn't take them off here. Probably he sleeps in them
and I can't blame him. The little trailer is never really
warm, only warmer than outside. We drink our coffee,
mumbling ritual conversations:

"How cold was it last night?"

"How cold will it be tomorrow?"

"When the hell is this goddam weather going to
break?"

"How much hay is left?"

And, "How are the critters holding up? It's getting
close to calving." John, Stan, and I know this weather
takes a toll on a pregnant cow, draining her strength as
she gives nutrition and weight to the unborn calf.

We haven't had a bright sky for a week; some days it's
a complete whiteout with snow drifting from the sky like
tiny down feathers, and other days it's spitting hard pellets
of ice. On these days, we cannot see a horizon or the floor

of the earth; we can only see dull gray or white all around us. Today, though, the sky is at least a pale blue, although the sun sits uselessly above us, offering no warmth. Ice crystals cling to the trees and shrubs—a beautiful sight, I suppose—but I'm not thinking about that.

When the coffeepot's drained, we climb back into coats, caps, and scarves. I have on all the warm clothes I own and still it won't be enough. Long underwear, wool shirt, wool sweater, Levis, and wool socks to start with. Then a vest under heavy coveralls, a billed wool cap with earflaps, leather mittens with wool liners, and snow boots with felt liners. I zip and snap and wrap and button, knowing the frigid air will creep in anyway and I will be numb before we finish feeding the hay.

Outside we go—no dawdling now.

We move to our routine: I halter the horses, Stan gets the harness, and John pours out the oats. Covered with white frost, the blonde-bellied Belgian mares, Pet and Maude, look like huge, fuzzy toys; they nicker at the rattle of the oat pan. While John curries them, I place the collars, lifting them around each horse's neck and shoulders and buckling them. Stan heaves the heavy harness up and we fasten the stiff leather straps. Bulky mittens make that almost impossible to do, but I know better than to take them off. Fingertips would freeze, touching metal buckles at thirty below.

When the team is harnessed we drive them to the sled. They step agreeably over the sled's tongue so we can hitch them into their accustomed places; we fasten the tugs and throw the lines up. John, Stan, and I awkwardly climb on, and the dogs jump on with us, wagging their tails as though this is just a jolly outing. John's dog Cactus has to have a boost, ancient as he is.

The horses pull the big wooden sled into the stack yard, with Stan driving to park snug against the haystack to make loading easier. I pull my wool muffler tight, covering my face, and we each start grabbing the seventy-pound bales.

No need to waste time deciding what to do, how to do it. It's the same each day: stack so many bales to a row, so many rows to a layer, one layer after another until the sled is loaded high. The horses stand in their traces, snitching bites of hay when there's any within reach.

When the sled is loaded high, Stan picks up the lines and clucks, "C'mon, giddup, Pet, Maude, giddup." The horses lean into the collars and the sled groans and creaks, inching into movement. The dogs have been scratching for mice as we load the bales, but now they trot out importantly, leading us toward the hungry cattle bunched and waiting. Their backs are humped against the brutal cold, and in a solid mass they press toward us, their hooves squeaking on the frozen ground. Each day they seem a little thinner and a little weaker, but we're giving them all the feed we can afford, making it last.

John drives first while Stan and I feed the bales, and we'll take turns—the driver has the coldest job as he stands still, holds the reins, and faces the breeze. Feeding is movement, at least, although hard work. Mittens make us clumsy, but we don't waste motion, and we fall into an awkward rhythm. We grip the bale strings with one hand, pull the bale near, slice the strings with a knife, shove the hay off in flakes with a knee, and reach for another bale, slow and steady. We keep our heads down, trying not to breathe the cold air. We pile the loose strings together as we work—no one wants them all over the feedground. Huffing and puffing, I pull the scarf over my face again as I step toward the front, taking the reins when it's my turn. My cheeks hurt, hurt from the cold. My eyes sting. I just keep on doing what I'm doing. I need to finish as quickly as possible.

Finally every bale of today's three-plus tons is fed. The cattle eat like they might never see another bite. We stop to let the horses blow; we tie the strings in a bundle and push scraps of hay off the sled. By now my face aches and my hands and feet throb. We turn back toward the corral.

Cold though we are, we can't hurry. Mistakes could cost money—maybe lives—in this weather. Stan uses the bar and ax to chop a waterhole open so the cattle can drink. John and I tend the horses. Habit helps us stay safe—we undo the tugs first, freeing the horses from the sled. (Remember, fasten the buckles from front to back when harnessing up, unbuckle from back to front to unharness.) First the tugs, snap them up, free the neck yoke so the wagon tongue can drop to the ground. Then the quarter straps, then the belly straps, then the hames straps. Stan flings all of that into the sled. Collars come off last, onto the sled also. We cover everything with a tarp in case it snows again tonight. (That's all we need!) John goes inside to make coffee. Stan and I turn the horses loose to feed and water. Finally we can head for the trailer, stamping our feet, clapping our hands together, slapping them on our legs to restore some warmth. We take the bridles inside—no one would put a cold metal bit in a horse's mouth at this temperature.

We swear at frozen buckles and weather and ranching in general; the trailer smells like wet dogs and dirty coats and overshoes. We drink strong black coffee while John smokes another "vitamin" and then we head for home.

We need to be at the bus stop at exactly the right time—the kids will need a ride. They will have homework, and they'll bring in wood to stoke the fire, and we'll check to make sure the old furnace is still working and that the faucets haven't frozen anywhere. We'll put together a hot supper. Stan will finish a few chores around the barn and plug the pickup in so it will start in the morning. Tomorrow's another day. We take our leave.

"So long, John. You'll be all right?" "See you tomorrow. Same time."

We kept those cattle in Basin until February 14 when the haystack was depleted. I will never forget the relief I felt, horseback and trailing those poor, thin cattle toward Shell Creek again. The old road left the Big Horn River, and the cows trudged across the back hills single file through the deep snow, as determined as we were to get home. They didn't falter; they seemed to know things would turn out right, and they pushed on.

On that same day, the weather broke—the Valentine's Day chinook—still in the record-books. As the weatherman had predicted, there was strong wind accompanied by eight more inches of heavy snow. The wind caused roads to close, shutting down schools and leaving people stranded throughout Big Horn County. Stan and I had asked some friends to help us that day, horseback, and John was with us, too. With our herd of cattle, we reached the highway just as the storm whipped up to fierce, and we had ten miles more to go. Big, hard snowflakes and gale winds reduced visibility to almost nothing, but we could not stop until we reached the ranch. The cows put their heads down and kept walking, their backs white with crusted snow. Someone in a pickup truck offered us a bottle of blackberry brandy as they passed—old John and I drank most of it. I was so tired, so relieved, and so happy to be home. By then, the Wyoming Highway Department had closed the roads, and our friends stayed overnight at the ranch.

The weather gradually warmed after the chinook, but we continued to feed what little hay we had at home until May when green grass finally emerged from beneath the ice and snow.

At the end of February my journal says temperatures reached 50 degrees. *Above* zero.

Looking back, we were grateful we'd gotten the cows home safely, and that Stan, John, and I hadn't frozen our fingers or toes, or worse, working as we had in that winter weather. The cattle were thin, but none had died. When the snowpack melted, it filled reservoirs on the range for spring and summer use, and the grass was abundant. Cattle prices came up a little, and we held our money together, at least breaking even and keeping the banker happy.

When people talk about disagreeable weather, I roll my eyes. "You should have been here in '78–'79. Now *that* was a winter."

8

Partners

On December 18, 2014, I wrote in my journal:

> We signed papers yesterday to buy Steve's property.
> Steve's body is "plumb wore out," he says. "Nothing's
> changed," I said. I hugged Steve and laid down the
> pen. Stan, Steve, and our son Tim shook hands, a little
> self-conscious at the formality. "We're partners, same as
> before." We had reached a tender, respectful agreement
> on the price for Steve's ranch. Our decades-long friend-
> ship and loyalty to each other is beyond any price.

We became partners at a party one night in 1984 when Stan
said to Steve, "I have a chance to lease some extra range. What if
we throw in together?" Steve was a cowboy, a relative newcomer to
the Big Horn Mountains, having moved his family to Wyoming in
the late 1970s, coming "up from Coll-arada," he said. "Too many
people down there." Steve, his wife, and two lively kids settled easily into Shell Valley. As a rancher, horse trader, and cattle dealer,
Steve somehow managed to exude steadiness and honesty. His
open smile and willingness to laugh at himself brought people to
him, in business and for pleasure.

"The way the market is," Stan continued, "I think we could buy
some old cows for short-term, calve them out, run them on the
leased ground, and then sell them fat in the fall. Or we could rent

out pasture by the day. I don't think Mary and I can do it by ourselves. But with a partner, I think it would work."

Stan leaned forward, describing the opportunity he saw.
Ranches sat vacant throughout Shell Valley during the early 1980s,
a common sorrow for ranchers and farmers all over America—the
national economic crisis created by the combination of high interest rates, low market value for livestock and crops, and inheritance
tax considerations.

Banks had foreclosed on property around us; many farms and
ranches were listed for sale. The bankers and realtors anticipated
an eventual return to property values, and they were eager to keep
ranches looking productive and attractive. They encouraged the
leases we suggested. Too, some of our neighbors sold their livestock and went to work in town, hanging on to ownership of their
land by earning steady paychecks elsewhere. "We can rent some
of those places for next-to-nothing," Stan explained. "We'll have
farmland, desert range, and mountain pasture."

Steve had told us when we met, "I only had one job in my life, and
I didn't like it. I decided I didn't want to work for somebody else,
ever again. And I never have." A partnership, though, appealed to
all of us. "Fifty-fifty partnership won't work. That won't be enough,
and we can't waste time doing the arithmetic. Each partner has to
give more than fifty. Partners, sixty-sixty—that will give us more
than 100 percent." We all agreed.

We guessed at the math of the cattle market and made rough cal-
culations on the bar napkins, and then we three made a handshake
deal. We didn't write by-laws or job descriptions, and we didn't hire
an attorney. We simply went to work. "Might as well be broke as
the way we are," Steve said, chuckling softly. "Maybe we can hold
on till things get better. It will give the kids something to do." Of
course, we knew that the kids—our four and Steve's two—would
work with us. "What choice have they got?" we said. In fact, the
kids called the partnership S&S, giggling and poking fun at Stan
and Steve, making wisecracks about "Slow and Slower."

On May 14, 1984, I wrote in my journal: "S&S is off and running. Our first *deal*. Looks like we're going to earn our money."

A friend, Jerry, from the nearby town of Manderson, had run short of grass, and he called to ask us to summer his eighty cow-calf pairs. He said he'd herd the cows from his farm to our gate at Sulphur, a ten-mile trail through the desert, and there we'd take official charge of them until fall.

Steve and our son Tim went to meet Jerry as arranged on May 13, but they saw no sign of the cattle so they came back home. Jerry called Steve that night, explaining that they'd gotten only partway. The next morning, Steve and I trailered our horses out to meet them. When we mounted up, we considered taking our lunch along, but Steve said, "Hell, they can't be far. They started *yesterday*."

We rode over one ridge and then another and another mostly at a hard trot, but we did not see them. Steve rode his favorite, Okey Paint, and I rode our mare, Buttons, both good travelers, and we kept going south until we topped a hill where we could actually see the green hayfields at Hyattville and Manderson. We stopped to let the horses blow, and Steve pulled out his old binoculars, squinting through the one good lens. "I see some dust. That's got to be them," he said. "Damn, they're a long way off, and I don't know how the hell they got *there,* but one of them's on a palomino, and I know Jerry's got the one I sold him a couple of years ago."

We hurried on to meet the bedraggled farmer-cowboys, finding them sunburned, sore, and worn out. They told us the cattle had looked at life on the range and headed back to the farm during the previous night. Jerry's crew gathered them and started again, this time with tired horses to push the bellering, balky cows—during record-hot temperatures to boot.

"Damn, we're glad to see you," one of the men said, mopping his ruddy face with a bandana. "Not sure we can get them there in this heat, and not a tree or a pond in sight." A big heavy fellow wearing a ball cap got off his ploddy horse to sit on the ground, leaning against a skinny little survey post just a few inches in diameter. "Damn, that shade feels good," he said.

We herded the cattle back to the trail and discovered they'd mixed with a neighbor's cattle. Steve and I decided to sort them at a gate ahead. Just then a motorbike came into sight, the neighbor himself riding full-throttle, rip-rip-ripping down the ridge. He hollered something we couldn't hear, but we guessed he was afraid we'd take his cattle with ours. He zoomed off the hill, but he misjudged the dirt and laid that cycle over—*plop*—right in the middle of the gate, just like a little kid on his first bicycle. He was mad *and* embarrassed then, but he waited while we cut the cattle out horseback before he rocketed off amid a cloud of dust and the rat-a-tat of the motorcycle. We started on again, several miles yet to go, the sun bearing down.

"I'll go get the truck and trailer," somebody in the Manderson crew said wearily. "It'll take me an hour or two, but at least we'll have some water and sandwiches and a ride home." When the truck returned, though, chugging around the two-track dirt road, we saw Jerry's dog lying in the cooler, panting happily in the melted ice. He had spilled whatever food he didn't eat. The tired, hungry cowboys loaded their horses and headed back toward home. Steve and I pushed the cattle on—the last mile.

We counted them through our gate at dark, congratulating ourselves on the first S&S business deal. At the truck, Steve and I dug out our lunch; we rummaged for the sandwiches we'd left there hours ago, and I poured coffee from the thermos. I raised my cup: "Here's to a good riding partner. That was a long day." Steve lit his cigarette and agreed. "Damned if not," he said.

At home, Stan fixed a stiff drink for us and we laughed together, reliving the day. "You never saw a happier dog!" Steve told Stan. "He was plumb tickled with himself."

Early on, an important piece for the partnership was to register a brand for the business. Our own cattle already wore the

Diamond Tail, and Steve's his K7, so we rebranded the cattle we bought with the Bracket C, a brand given to us by our old friend Bill Clifton years ago. Any rented-pasture livestock wore their owners' brands, and every single critter must be identified and accounted for at the end of each season.

At one time, we partners ran livestock belonging to nine different owners, in addition to our own. We got damned good at sorting cattle and reading brands, and we liked riding together. We separated cattle constantly to meet their owners' schedules: "Dale Aagard wants his cattle on Tuesday. He's ready to go to his forest permit, and he'll send a truck for them. Mullins will help us gather theirs on Friday; their grass at home is ready. Drwenski can take his cattle now; we'll trail them around from the Saban lease to the head of Trapper Canyon." Some of the leased ranches had been neglected long enough that the fences were in poor shape, so we regularly answered phone calls telling us the cattle were out, or in the wrong place.

Our workplace stretched for miles—a spectacularly beautiful jigsaw puzzle of land, nearly a 100,000 acres, which lay far, far south of Shell Valley into the desert land beyond Red Mountain and the Mail Trail—almost to Hyattville and Manderson. From there it stretched north to the mountain rims of Bear Creek and Beaver Creek, then from Shell Valley to Granite Pass, Black Mountain, Bald Ridge, Trapper Canyon, and Trapper Creek. I learned the workings of each new setting, and I treasured every moment, finding crevices and crannies, hidden springs and reservoirs. I learned to trust each landscape and the wisdom of old-time trails that usually lead to water.

Beyond the cowboy work, we three partners used our different abilities to get things done. Stan dealt with the landowner contracts, the Bureau of Land Management (BLM), and the United States Forest Service (USFS) for permits and paperwork, and he knew the maps, landscapes, trails, and waterholes on most of the places we leased. Steve did most of the buying and selling in cattle deals. I kept track of the money, did the billing, packed coolers, and filled in everywhere else.

Each lease required meticulous record-keeping: cattle numbers, days of grazing, rent figures, and our own expenses. Steve copied figures from his shirt-pocket book into his desk-book every night: cattle counts, pasture changes, vehicle trips, and mileage. Stan and I did the same, double-checking each other at the end of every month. "I've got it right here," Steve would say, thumbing through his book. "Seven trips to Red Basin for you, nine for me this time," and we tallied each other $50 credit per vehicle trip, for fuel and wear-and-tear, equaling out the days each crew worked. We never had a serious disagreement about money, not one.

Each of us listened hard everywhere we went, hoping to hear of an opportunity. "Talk about three vultures," I said. "Just waiting for somebody's misfortune." "Trying to prevent our own," Stan countered. "No shame in that."

Some of what Steve called "bargains" barely hobbled out of the truck. "The price was right, so I took a chance," he said. "We can make a little money on them when they're fat." He gambled on cattle with their ribs showing, lame cows, and motley calves, each time saying with a grin, "You know, we can't afford the top-drawer stuff. These will be all right by fall. Nothin' wrong with that cow. She's got a little experience, is all." The kids laughed and laughed. "Give me your tired, your poor, your huddled masses? Did she write that about S&S?"

No deal too big or too small, we thought. "I talked to a guy," Steve said. "He's got a little package of yearlings he wants us to summer. I told him we'd do it." Or, "Fellow at Worland wants to sell some young cows. Says he's tired of fighting the work of it. Just wants to get rid of 'em." And, "Man, we hit a lick today. That new guy called—named a price for his cows. He didn't last long, ranching."

Steve showed a handsome-cowboy look when he put on his good Stetson and nice jacket, but mostly he wore baggy, faded Levi's, boots run down at the heels, and a sweat-stained cowboy hat; later, suspenders across a worn denim shirt. He claimed that looking too good would scare people away, and sometimes that was true. Meeting Steve for the first time, a fellow in one of the trader-deals said to Steve, "I think we'll get along. There just ain't nothin' too stylish about you."

Steve bought some big-framed Charolais, white farm cattle from all the way into the Missouri hill country. He flew there to make the deal and rode back to Shell with the trucker, proud as punch to tell us of his shrewd bargaining. We didn't realize the cows had never been handled horseback and they were scared to death of horses; they simply bolted if anyone rode near them. They were comfortable in the fenced pasture where Steve unloaded them at Bear Creek—one of the ranches we leased from the bank—but when we opened the gate to begin the trail to high summer grass, they took off at a trot and we cowboys couldn't slow them down. We worried they might drop dead in their tracks as the day heated up. We finally placed cowboys where the cows could barely see them, and in that way we lightly steered them up the mountain, some thirty miles over two days. When the fence stopped them, we put the cattle through the gate and didn't disturb them until fall; by then they had learned how to live "outside" and they were easy to handle. We teased Steve all summer about the "Missouri Trotters," but he had the last laugh because they gained weight, raised good calves, and made money for us.

"Man's got a couple of horses he'll pay the kids to ride, too," Steve said, "but let's be careful on those, especially town horses. Usually it's some spoiled damned thing that belongs on a French-man's plate." Beyond his uncanny eye for quality and soundness, Steve was a horse whisperer in his own right, and he knew a million tricks to fix bad horse habits. I saw him use foot ropes and stock whips as teaching tools, as horse clinicians and trainers do nowadays. He never abused a horse, only corrected it and patiently taught it how to behave. "Usually it's the rider who needs fixing," he said. "And if somebody says it's a 'cowboy's horse' just stay away from it," he told the kids. "That's horse-sale language for 'nothing but trouble.' Remember, everything's fair in a horse trade, and we can't have anybody getting hurt."

Our rowdy crew of kids, Carol, Tim, Sara, and Dan—college, high school, and junior high—and Steve's two, Cammy and Cody—junior high school—worked well together, too. They were all good riders who could handle livestock, change tires, manage jumper cables and horse trailers, fix fence, and drive our pickups. As Stan told the man who questioned Cody's age, "Oh, he's old enough to drive. Just not old enough to have a license."

The rancher life of no money and too much work took its toll on many marriages of that day, and Steve's was one of those. He and his wife divorced, and Cody and Cammy remained with him. He struggled to keep up with single-dad tasks plus the ranch work. Stan and I shared that load as best we could, feeding everybody at our table often and herding kids everywhere. My journal tells that they all pitched in at home: "Sara had supper ready that night." Or, "Kids cleaned up the kitchen, did the laundry." "Boys put the horses away and corralled fresh ones for tomorrow." They bickered over the sandwiches in the cooler and did their share of petty squabbling, but all of them held tough during long, long work days.

For Steve, Stan, and me, part of each day's job was getting the grumpy, tired kids out of bed before daylight, fixing breakfast, saddling horses, packing lunches—then loading horses, kids, and dogs for an hour's truck and trailer ride. On one day we planned to meet for an early start, knowing the desert country would be scorching hot by mid-day. At the prearranged spot on Red Mountain, we found Steve's rig there ahead of ours, but in the dim light there was no movement or sound, and we correctly figured the little crew was catching some sleep while they waited for us. After a long moment of peering through the windshield, someone in our truck said hesitantly, "On top of Steve's trailer? Is that a . . . chicken?" A *chicken?*

"Hell," Stan said, "it *is* a chicken!" The kids bailed out of the truck yelling, "Steve, Cody, everybody! Wake up! It's a chicken!" Steve and his family tumbled out, confused and groggy. "A chicken? Where the hell did she come from?" She had roosted atop the trailer at Steve's place, clung there for twenty miles over a bumpy dirt road, and now took a halfhearted flight into the sagebrush with kids and dogs in hot pursuit. We all joined the chase, running, tripping, falling, laughing—forgetting about the "early-start by God" riding orders. When we finally got the old white hen captured, we left her in the horse trailer all day while we worked, and then we brought her back to the ranch where she clucked around indignantly for weeks, her tail feathers comically fewer after each encounter with the cow dog. "She's not fainthearted, I'll say that for her," Stan said. After a while we took her back to her rightful home at Steve's, but the "Cowboy Poultry Gathering" is a favorite story.

Eventually Steve remarried. His new wife, Margo, brought her three kids to join our work force. The ranch work created a revolving door for other youngsters, too. Some came from troubled lives or circumstances; others just wanted something to do and to be where there was action and fun, where every day brought pranks and teasing.

One day we filled a bottle with water to carry horseback to the cowboys. Cammy reached for it eagerly, as the label said it was the new power drink. "Oh, I've been *dying* to try that. Dad *never* buys good stuff for us," she said, taking a swig, then another. "Phooey. It's not that great. Tastes just like water to me." She handed it to Cody. "See what you think." He drank, spit, and said, "Well, no wonder, Homer. It *is* water!"

Steve's family members affectionately called each other "Homer," an allusion to mistakes or, as they said, "dumb attacks." Soon we were all saying it. "Geez, Homer, your pants are on fire!" And they were! The baler twine Steve stuffed into his back pocket had caught fire when he backed up to the stove.

"Having any luck there, Homer?" Stan was trying to teach his red heeler a "lesson." The dog kept crawling away just out of reach under the pickup, while Stan squirmed on his belly in chaps, boots, and spurs, dragging through the dust saying, "Come on Tuffy, nice doggie, nice doggie, I'll kill you, you little sonofabitch, when I get my hands on you." He never caught the dog that day, and the S&S crew still laughs about Stan's dog-trainer techniques.

The youngsters who stayed, worked hard—long, long days. "You'll have to come get him," I told one boy's mother. "I thought he could stay for the summer," she said. She asked if we'd reconsider, but I had to say, "No. He doesn't want to learn. He won't try. I'm sorry. We have work to do." Age was not much of a consideration with the days we faced. Everybody worked.

One morning we sent Carol and Cammy to Red Basin, a huge bowl of sagebrush and grass that sits off the east rim of a five-mile incline at Red Mountain. They drove a pickup and trailer loaded with a couple of saddle horses and some yearling heifers that they were to deliver to a far pasture. I knew something was wrong when the girls drove the pickup back down our lane too soon, without the trailer. Carol's voice shook as she told us how she tried to slow down when she reached Red Mountain's steep downhill S turn, but "the pickup brakes didn't work; the trailer fishtailed, and it rolled off the embankment and tipped over." Cammy said, "We got the horses out and the cattle too; I think they're all right."

"What the hell?" Stan said roughly. "You know better! Going too fast—probably playing with the goddam radio dial!"

"I tried to shift down," Carol said, near tears, "but I couldn't get the right gear." "I don't know what gear the *truck* was in," Stan raged, "but I can guaran-goddam-tee your brain was in *neutral!*"

Steve stepped in, hands on his hips, hat pushed back. He said gruffly to Stan, "Now wait a minute, Homer. You need to say out loud: 'We're just damned lucky nobody was hurt.' And that trailer, not to mention the pickup, is a piece of junk, and expecting them to drive the damned thing was just plain stupid." Stan took a deep breath and admitted Steve was right, but the moment had passed to comfort the girls. The men took Carol and Cammy back to Red Basin where the horses were standing unhurt, tied where the girls had left them. But the yearlings were gone. They had taken off at a high run, the girls said, and we never did find one of them. Eventually we all could laugh about the incident, and the girls said that maybe the heifer with the broken horn had run clear to Montana.

Nobody laughed, though, when a $60 bottle of livestock medicine slipped out of my hand and broke. Nobody laughed if someone lost a halter or a coat that would have to be replaced, or if somebody had a dumb attack and put some livestock in the wrong place, or forgot to gas up the truck. Because, of course, time was money.

Our young crew learned Life Lessons: "Pay attention, to everything," we told them. "Some animals die and some live, and it might be up to you. Some things you can handle, some you can't, so learn the difference. Be careful. Watch out for each other. Do what you say you will, and what you're told. Ask for help when you need it. It can mean life or death. Always pay attention."

Despite plenty of close calls, no one was ever seriously injured or truly lost. "I knew where *I* was," Dan said. "I just didn't know where it *was,* that I was." The worst injury was a broken arm, although horses ran away or fell, and people got bucked off. One boy's horse fell into a ravine, wedging him so that it took lariat ropes and several cowboys to get him out—but neither the horse nor boy were hurt.

Tim, horseback, once tried to rope a high-headed black cow, thinking he'd drag her into a trailer. "Leave her alone," Stan called. "Get out of there, right now. She's *mad*—she'll hurt somebody." Tim, a cocky teenager, reluctantly left the corral just before the cow went crazy and jumped into the fence rail, breaking every-thing in her way. "Watch out for that kind. You've got to know what you're looking at."

The work never stopped. Stan had a big game outfitting busi-ness, too, which geared up in the fall and kept him busy guiding nonresident elk hunters. It brought in good money but left me and Steve working alone after the kids were back in school. An early storm hit the mountain one October in the mid-1980s, bringing first a foot of snow and then sunshine and cold again to create a miles-wide white crust covering the mountain pastures: Bald Ridge, White Creek, Hazel Early, and Mackey. The cattle put their heads down and waded through the snow as they tried to get down to the valley, breaking some fences and walking over others. Sev-eral neighbors had livestock near ours, and the cattle mixed into one enormous herd with different brands and owners. Steve and the neighbors and I rode horseback to find the cattle that could not be reached by road or trail in the frozen snowpack. I wore so many warm clothes that I could barely get on or off my horse.

The gather took several days. Steve and I stayed close to the work, on the mountain at the White Creek camp with Stan and the hunters, where we could share the cook and bunkhouses and corrals for our horses. We left and returned in pitch dark, meeting the other cowboys each day. At last we bunched the cattle into the neighbor's mountain pasture, and then we took them to the historic Wire Corral. The following day we took them down the icy Beef Slide to the foot of the mountain, onto Trapper Creek, and we left the cattle there until we could sort them the next morning. We had trucks there to take us home with our horses, but when some hunters offered us free drinks at the nearby Five Sisters, a roadside bar, we accepted. They wondered if we'd seen any elk, but we hadn't—and they kept buying us drinks and asking us why anybody would want to be a rancher. We thought maybe if we drank enough we could answer the question.

We stayed out late, so I was tired and grumpy the next morning when I met the other cowboys at Trapper Creek. Horseback we began to sort for ownership, planning to trail our cattle home in separate herds, in different directions. Things worked smoothly until two men pulled up in a truck, ready to claim their cattle. We had called them earlier in the week when we really needed help, but they didn't show up.

Steve and the others muttered, but no one said anything out loud when the men stepped out of their truck in rubber boots, striding into the icy pen, carrying cattle prods. They poked their sticks around, moving cattle here and there. I was afraid a horse might slip and fall as they spooked, jumping away from the waving arms and prod sticks. After a few minutes of trying to work this way, I blurted out, "Since you didn't bring horses, just wait in your truck."

Things stopped, dead quiet—nobody said a word. Nobody moved until the men afoot climbed out of the corral, sticks and all. We finished sorting horseback, then we loaded their cattle for them and they drove away.

Our cowboy crew headed to the Five Sisters again that night—finished, ready to celebrate! Our week's work got a lot funnier in the retelling, even the part where we and our cattle were padlocked *inside* a pasture. The sign on the gate said, "NO TRESPASSING! Absolutely No Hunting! This means YOU. No In-laws Outlaws Friends or Neighbors." We laughed about frozen sandwiches and droopy coveralls, making hilarious fun of each misfortune. Then one of our by-now rowdy cowboys said, "Well, Mary sure put the run to them sharp-stick boys! She told 'em, 'Get out of the corral if you ain't mounted,' and by God they did! Where the hell was they when we needed them, anyway? A bunch of farmers, that's what." We laughed and laughed, "Yeah, the sharp-stick boys!" The barroom boys teased me about being so bossy, but they admitted I was right. "*Now's* a fine time to speak up," I said. "Thanks."

The lease properties did eventually sell to others and our opportunities waned. By then, though, we had gained the financial foothold we needed.

April 29, 2015:

Steve's auction, a rainy, cold day. Steve peddled his left-over equipment and "anty-ques," and all description of saddles, wagons, harnesses, piles of junk which accumulate on a ranch. He would have no need of those things now, and most of them haven't been seen or used for years, anyway.

A big crowd showed up for the sale, eager to find bargains. The auctioneer worked his way along, row by row taking bids for every item and making jokes. He, too, has known Steve a long time.

A smaller crowd collected at the coffeepot where the conversations turned toward "the S&S years." Rain dripped off Steve's old cowboy hat as he joined us; his ragged yellow slicker hung open and, like always, Steve laughed at himself. "Might be able to afford a new slicker at the end of the day."

His "plumb wore out" body gives its history of ranch work—bad knees and shoulders and stiff hands from lifting, shoveling, walking, bending, and cowboying. He was struck by lightning once, had several horse wrecks, punctured a lung, broke some ribs, and we tease him, recalling his adventures. He laughs, telling on himself, "Old Dawson's damned house-pet of a horse Roy-Boy got away and run clear off to the Ranger Station. I could have killed him. And that grey of Walton's, when he tossed me I hit the ground like a great big watermelon! Now that coulda been bad!"

Stories multiplied as we reminisced and, despite the drizzling rain, we stood to laugh and listen. "I just thought I'd teach that old cow a lesson and she put me right up on the haystack. I damned near didn't get up there!"

"Remember that crazy old irrigator we had at Beaver Creek, when he called us to come up there in the dark? He said we were being invaded! He was damned excited, but all it was, was headlights from a trail ride bunch getting started."

"Remember, he thought Pat's fancy sunglasses were hearing aids!"

"Stan's tire chains—remember if you got stuck, you had to tie them with baling string and bungee straps?" "How about when the buyer asked the horse's name, and Pat said, 'I call him Alpo, because he'll make good dogfood.'" "And Cody's horse Fred—Frederic Remington—that big jug-head with the wild eye?" "Stan's 3-wheel drive truck that was supposed to be 4-wheel drive!"

A little sadness crept up on our little group as we acknowledged the camaraderie we had shared. "I learned so much, those S&S years," a young man said. "I was just a kid. I didn't appreciate what we had. I wish my own family could have that experience." Another said, "Is it too late to say thanks?"

"Hell, nowadays they'd give us a grant or some damned thing, for keeping all of you off the street," Steve laughed.

"It feels good," someone said to me, "that Steve's place is still in the family. I mean, you're still partners, aren't you?"

"Yeah, still partners."

And we have no regrets, except that Steve died in 2016. Our partnership was a sweet, enduring friendship, a privilege and an honor for us all.

9

A Ranch Divorce

When I heard that our friends from Johnson County had brought their band of sheep over to the Big Horn Basin for the winter and that they would bunk in a shop near a rented pasture, I called to find out how they'd be spending Christmas. "Will you be going home, or would you like to come to our place for Christmas dinner?"

"We'd like to come!" Bob said, then paused and took a deep breath. He chuckled. I visualized his ruddy face—his dark eyes and jolly big smile. "Our family dinners haven't been all that much fun lately. You know—the ranch divorce. We haven't got that right, yet." He told me he and his brothers had recently divided the long-time family holdings, and he explained that they had some "rough spots." "I know what you mean," I sighed. "Please come."

A ranch divorce—not man-and-wife but instead the permanent division of land, livestock, and, too often, a family. Someone will take custody of hard-used property and perhaps a debt. Two halves will be less than a whole, probably. No matter what the offer or the settlement, there will likely be resentment and hurt within those families. Some will assign blame; some will feel a sense of failure. All will hope for harmony and approval.

Bob and his wife did join us for a grand Christmas dinner, sharing laughs and reminiscences interspersed with topics of weather, livestock markets and grass, and family and statewide gossip. Finally we relaxed with a glass of brandy, and Bob leaned away from the table, ready to talk. He knew we had been part of a ranch division, separating our share of a family ranch from Stan's brother in years past.

"So, how'd it go in the long run, your ranch split-up?" Bob asked. "Was it worth it, that divorce, hard feelings and all? Did things turn out right, for you?"

"Yes." Stan and I both nodded carefully, slowly. "It was the right thing to do. But it took a long time, really, for the scars to heal. We split the place more than three decades ago, and a lot of hurt and hard feelings stayed around. I think that's mostly behind us, now."

Stan and I joined his family's ranch business when we married in 1963. Of course, we didn't foresee a divorce; "live happily ever after" seemed in the offing. Stan, his brother, and his father—three handsome rancher look-alikes—set about solidifying and expanding the sheep, cattle, and farming enterprises that were in place at the "home ranch." Stan and I were delighted to find a welcoming atmosphere within his family. Certainly there was plenty of work for all, and the partnership seemed ideal.

Stan gradually took responsibility for the livestock, while his brother's interests leaned toward farmland, crops, and equipment. Their father, Howard—active and fit—enjoyed the company of his sons, and he loved the ranch he had put together.

In the early years of the relationship, the men talked on the phone every day, enthusiastically exchanging ideas and making work plans and decisions. Our three families had fun together—wives and moms shared coffee, exchanged favors, and sometimes met for late-afternoon cocktails while the children frolicked and teased each other, playing on the lawn or beneath our feet. Other days, the collective seven kids romped the miles between our houses. We let them run together in a rowdy, noisy pack—they galloped their horses everywhere they could, or they waded in the creek and played in the barn. The girls practiced hairstyles and makeup while the boys shot BB guns, rode bikes, and climbed hills. Friends forever, they thought.

Stan and I taught all of the children to ride horses, and we hauled them along with us to cow camp, 4-H meetings, and school activities as though we were one family. A *National Geographic Magazine* photographer caught one image from those times at a County Fair: our twelve-year-old nephew in a cowboy hat, nervously trying to manhandle his big 4-H show steer. Stan's hand barely shows in the photo, but it's there on the boy's shoulder, offering reassurance and guidance. Stan and I loved the nieces, too, and we sat faithfully in the front row to cheer for them at school musicals, high school pageants, horse shows, and queen contests.

Gradually, and perhaps inevitably, the one-big-happy-family became less than happy. "Part of our trouble was money, or at least that's how it began," we told Bob. "This was the late 1970s, remember, and the livestock business was seeing record-low market prices. There wasn't enough income to meet ranch costs. The ranch paid us meager wages, but they didn't reach far enough to save for kids' college educations or pay for living expenses like health insurance and groceries. We realized we had to figure out a future for all of us—especially Stan's parents, by then in their late seventies." Stan's father, Howard, said again and again, "Things have to turn around. I hope we can make it till things turn around." I still remember how he frowned and sadly shook his head. The markets didn't turn around, though, and the men nervously sought bigger loans to cover the cost of doing business. The banks were reluctant. "Your balance sheet isn't good," the bankers said. "You need to get a grip on your debt."

The three partners differed about priorities for money and time management. These men were frustrated; they didn't talk much, avoiding complicated discussions about long-term plans. No wonder, since they could not agree on topics as simple as shearing dates, shipping, or harvest plans. Anger erupted around little things—someone forgot to call the brand inspector, or someone opened the wrong gate. Someone used the wrong saddle or misplaced some tools. As families we pointed fingers at each other, finding excuses and looking for someone to blame. Operating costs kept climbing—gasoline, machinery repairs, and utilities—and the work became more and more disorganized.

We wives pretended to be oblivious to the men's disagreements, although it became awkward to find topics for safe, easy conversations. Naturally, each of us stayed loyal to our own husband's point of view. My own ranch upbringing gave me some awareness of our business plight, but as was normal for most ranches of that time, the corporate structure did not include women, and we wives had no real voice, even had we known what to suggest.

Stan's brother was a passionate, articulate spokesman for the complicated issues of the times—he traveled extensively, sincerely representing agricultural causes. Stan and I complained about his frequent absences from the ranch, even as we respected that dedication to national issues. Without communication, without a functioning partnership and division of work, everything suffered—the ranch crops, the quality of livestock and rangeland—and family ties frayed, too.

Our families had always shared traditional family dinners and parties, and for a while we tried to continue. "They became sad affairs," I told Bob. "We used our lovely crystal and china, and we fixed delicious food; we gathered together, dressed for the occasion, and everyone crowded around the table. Gradually, the once-festive occasions turned into ordeals where nobody joked or laughed; we politely passed food and compliments, but we didn't linger over coffee and dessert." In easier times, we women had exchanged recipes and food ideas and laughed our way through clean-up camaraderie in the kitchen, but when someone said,

"Let's just use paper plates next time," we stopped suggesting a "next" family dinner. The children seemed puzzled, but some were teenagers by then and, being teenagers, they showed little enthusiasm for family events.

Indeed, not all that much fun, as Bob said. The dinners were a small reflection of our state of affairs—Howard, as the patriarch, blamed himself for letting finances and personal issues get beyond control. Yet disappointing as his own failure must have been to him, ranches all over Wyoming and the West were in similar positions.

"It seemed that if we split the property, one or both of the brothers might have a chance for success and maybe also offer financial security for Stan's parents," we told Bob. "The way it was, together, we didn't have a chance. So, a divorce, as you say."

With the help of an estate-planning specialist, the family decided upon an IRS-approved liquidation, which required that documents be meticulously prepared, then signed within a given thirty days to transfer the assets. Stan and I spent afternoons of that cold 1978–79 winter at the Big Horn County Courthouse researching old titles and deeds, digging out maps, surveys, and documents for water rights and federal land grazing permits. All property would have to be identified and then divided equally and fairly.

"We were *all* sad, really. Disappointed, disillusioned—and scared. Stan and I had four kids, bills to pay, and expenses everywhere we turned. Of course, we felt a sense of failure and dishonor, and as I look back now, I suppose each family felt some kind of rejection."

In the end, we faced the facts: Stan and his brother would each assume a huge debt in addition to land, livestock, and some worn-out machinery. Each would make scheduled monetary payments to Howard and Maureen and to the two off-ranch siblings who, Howard believed, deserved "something of the family holdings." Understandably, off-ranch family loyalty to the ranch did not extend to the debt that had accumulated.

The terms of the "divorce" were clear. Lawyers say they hate divorces: millions of dollars can be divided easily enough, they say, but when it comes to figuring out who gets the cat and the dog, all hell breaks loose. It was true for us, too. Decisions about the major assets— houses, fences, cropland, and rangeland—seemed somewhat logical. Stan and I kept the original ranch headquarters and the old log home where we lived, as well as mountain land at White Creek, and our share of farmland. Similarly, Stan's brother's half of the property included excellent farmland, along with an expanse of good mountain grazing land at Bald Ridge, as nearly equal in value as could be managed, with the guidance of an attorney and an accountant. Even so, those trade-offs were fraught with tension, apprehension, and fear as we abandoned the concept of "safety in numbers," leaving the partnership behind.

Our fiercest "dog-and-cat fight" was over the historic brand, the Diamond Tail, used by the family since 1906—an iconic object of sentiment as well as a business tool. Each brother felt he deserved to inherit that nod of respect. Howard agonized; finally he left the brand attached to its original property—Stan's part of the ranch. Dividing the saddle horses brought more hell, although in the big picture their monetary value was insignificant. Each of us had favorites, and after the sheep and cattle were counted out and separated, the two brothers met at the corral one chilly morning; they tossed a coin to determine first pick of the horses. From the fence rail they called out in turn: "Nibs." "The buckskin." "Snoopy." Each choice brought a new sorrow, although the deal was fair and square. "I felt like I'd lost a friend," Stan said of one. "That horse must have packed me a thousand miles." Each child kept one trusty favorite, the only bow to sentiment.

Next, the men divided equipment and machinery. Most of it was beat-up and old, but each half-ranch needed fundamentals—a truck, a baler, a tractor. Weeks ahead of time, Stan and I scribbled lists of what we'd choose first, then next, then we'd rearrange the list. Which did we need more, a tractor or a baler? A stock truck? Finally, the men sat at Howard's desk, lists in hand, alternating selections. They didn't chat; neither offered to share or compromise. They did the math, they copied the final outcomes, and they left the room, each as owner of his own ranch, "to sink or swim."

Around us other ranches struggled, bringing a special vocabulary for the times: "Going under," neighbors whispered. "In Trouble." "Have you heard? Bank took him down." "Selling out." We hoped not to be among the unfortunates. Time passed. Our families and our ranches headed in different directions. Kids grew up, went to college, and married. Their loyalties tumbled in confusion, and their bonds visibly unraveled; the cousins who were once best friends stepped apart.

"What would you do differently," Bob asked that day at Christmas dinner, "if you had a chance?" For me, it was an easy answer: "I'm sorry we never talked to the kids, the young cousins," I told him. "They never heard both sides of things, and we never heard their questions. I think they all felt betrayed. I'm sorry for that."

"For myself," Stan said, "I loved the land. I regretted that I couldn't take care of it, couldn't keep it. It had nothing to do with greed, just that I felt protective, responsible, obligated to look after it. And I didn't want to part with any of it. That was hard for me, and it still is, some days. It was such a beautiful ranch, all together."

He shrugged, then smiled, suddenly self-conscious. "They're not making any more land, right?"

"What about *your* family," Bob asked me then. "They're ranchers too, aren't they?" I answered, "Yes, but we were three girls. Two of us married ranchers ourselves and had already stepped away, leaving our younger sister and her husband as the only partners with Dad." As Stan's father had done, and as many rancher-fathers want to do, Dad determined there should be "something" in the family assets for those of us who had left the ranch, including his own sister, a widow. Dad clearly defined a monetary gift for my older sister and me, and one for our Aunt Helen, although it was a financial burden for that ranch. "We said 'thank you,' I told Bob, "and gratefully accepted what we were offered, no questions asked, leaving that ranch—and our relationships—intact."

"Before we knew it," I said, "a generation or two passed. Stan and I and our grown-up family made choices, too—sad partings, different life directions, but not divorce. Three of our four children stepped away, allowing the business to develop strength and unity. I hated that, but they have made good lives for themselves in other places. We're still a family, and our ranch is still whole. So I guess we did the right thing."

All of us at the Christmas table nodded, quiet for a moment. One of us raised a glass, "A toast: to the future of a ranch. To its character, and to its presence in a landscape, and to its people, who care about it."

"Hear, hear," I said, smiling, ready for a lighter mood. "Think about this! I remember like it was yesterday, when Stan and I trailed our cattle to the range that first spring—our *own* cattle. We knew we'd have to pull ourselves out of the hole; we knew we'd make mistakes, but they'd be *our* mistakes. I remember how proud we were, sitting horseback after we turned the cattle loose, and they spread themselves across the hills. I feel the same satisfaction today. Good luck, Bob. Hear, hear! Here's to a fresh start. You'll find your way."

10

Deep Tracks

September 6, 1989:

Mike died last night. She'd been in the hospital just a
couple of days, can't believe it. I don't know how I'll
live here without her.

The first time I saw her I wanted to meet her, the woman standing
beside a counter at the Shell Store, talking easily with the owner, a
paper bag of groceries hiked against her hip. She was lovely, simply elegant in faded Levi's held by a soft leather belt and engraved
buckle, and on her feet, beaded moccasins. She wore her hair in a
heavy single braid—graying then, and silver later. I did meet her
at last, when we both signed up for a night class in town, a college-
level class in finance. She was there because she had "cabin fever,"
she said, and I guess I did, too; I just didn't know the name for it.

She lived above Shell on Horse Creek, so we drove together to
the school that winter, visiting carefully at first and slowly learning about each other—at first simply that we were two ranch girls
transplanted by marriage to farm country. She had married at age
thirty-two, to a mining executive, Frank Hinckley, a "dirt stiff" he
called himself. He proudly told the world he bought Mike a little
ranch below the "W"—a geologic landmark against the Big Horn
Mountains—to keep her happy. They had three young kids, and
Stan and I had our four, so we had much in common.

She explained her nickname: "When my mother was pregnant, she and my father lived in the Powder River country near Kaycee, and wherever they drove they had to stop for gates. Dad told Mom not to worry, that 'when Mike gets here he can open the gates.'" The baby was a girl, but they called her Mike anyway. She grew up well schooled in cowboy ways, tutored by her family and the hands at the TTT Ranch (commonly known as the Three T) where they lived. As we got acquainted, I learned that "in her day" she was the University of Wyoming's first official Rodeo Queen. Mike and her husband, Frank, liked the Shell Valley and Horse Creek, but she felt "kind of housebound," she said. Opportunities for riding and working cattle seemed scarce here in her new setting.

One afternoon, Mike invited me to her house to study, and I remember my astonishment when I saw the sterling silver tea set and candelabra placed casually beside an old revolver. I learned that her full name was Mary Frances Tisdale Hinckley, and that her heritage included characters prominent in Wyoming's rangeland history. She explained that the pistol—a Colt Peacemaker—was a relic of the Johnson County War, and she told me of her grandfather, John A. Tisdale, who was dry-gulched, shot to death at the outset of that conflict in 1891. Her hand-tooled bridle with its graceful, narrow reins and silver bit were draped over the arm of a chair, along with her custom-made spurs.

Mike found her niche at local brandings, once her roping skills were recognized. The invitations had been slow in coming to a newcomer to the area, but before her marriage, she had stepped into position as the cow foreman at the TTT Ranch during World War II, making her a legend on the east side of the Big Horn Mountains. As she became better acquainted with Shell area ranchers, she was a highly sought-after helper for any kind of cattle work. Because of Mike's participation and her presence, it became acceptable and respectable—and fun—for me to bring out my saddle, taking a place horseback among the men of my husband's family, who weren't used to women cowboys.

Both Mike and I felt displaced sometimes in a community of gardeners, bridge players, and luncheon-goers, and we looked for reasons to ride together. She knew the same ways I knew—trailing cattle slow and easy, working a herd carefully and meticulously. She was "a damned good hand," to use her own highest compliment. For the simple joy of it we raced our horses, complained about the boss (whoever it was), shaded up at noon, and smoked cigarettes, talking and laughing. We shared stories about ranch country and cowboys. She told me how it horrified the "green-eyed wives" when she traveled to Omaha on the train with the cowboys during World War II, to ship the TTT cattle to market. She mentored my children, too, praising the things they did right: "Did you see little Sara, making a hand—she and old Peanuts getting to the low side of the herd without being told."

We created new stories, too. Mike never left home without her lipstick. We used it for a pen once on a piece of cardboard, to explain to Stan why the truck was where it was, and why we were somewhere else: "Brakes broke. Us Copeman's Tomb after bulls."

One evening the two of us pulled in at the roadside bar and parked our truck and trailer outside, thinking we deserved an icy cold daiquiri on our way home from the mountain. We ordered at the bar before we realized that neither of us had a cent to pay our bill. A group of well-dressed men stepped in to the bar just then, clearly there on business. One of them greeted us. "My gosh, a surprise to see you girls here. Out on your own? Let me buy you a drink!" We accepted, of course, and laughed and laughed at the close call, saying, "Good old Ken! What would we have done?" Another time we stopped at the bar and had enough money to stay too long. We didn't realize that Stan came to get the truck (loaded with our patiently waiting horses) and took it home. Then he called Frank, and they showed up together to join us for a night of carousing.

Mike kept her own string of well-trained, well-bred quarter horses. Apart from the family budget she had a little bank account she called the "Quarter Horse Fund." "That Quarter Horse Fund is the damnedest thing," Frank said. "I don't know where it comes from, but she's always got money in it when she wants something." She named her favorite horses for places and rivers she'd ridden: Chuggie, for Chugwater; her sorrel, Powder River; and her fancy bay mare, Crazy Woman.

Mike and I shared a liking for good horses and good books, and we both loved pretty clothes; we exchanged recipes and commiserated about the cost of groceries and about family life, the way friends do. "The trouble with kids," she said, "is they're just so *damned steady*. Day in and day out." Sometimes Mike phoned to say, "Let's run away." We might drive to Sheridan to shop at King's Saddlery or have lunch at the Elks club, or go to a horse sale, or maybe throw our saddles onto horses and go to ride somewhere new for an afternoon. Best of all, we liked cowboying together.

"Deep tracks." The phrase pops into my mind and I laugh, thinking of Mike, and remembering our cowboy days.

The Padlock Ranch is a large, prestigious, corporate ranch on the east side of the Big Horn Mountains, established long ago in Wyoming and Montana. Then, as now, the Padlock Ranch hired cowboys whose only job was to look after cattle. Many were rowdy young buckaroos who regarded themselves as true professionals, riding in style with ropes coiled and ready to throw, big-roweled spurs, wide-brimmed hats, chaps, and wild rags. Some were good hands; others were complete gunsels, in the old vernacular, and some were noticeably condescending toward the mom-and-pop outfits like ours from the lowlands of Shell Valley. After all, our cowboy work was mixed with such ranch jobs as putting up hay and fixing fences. To my cowgirl friend Mike and me, those fellows seemed self-important and a little ridiculous, while as *women* in their cowboy world, we must have seemed like hard scratches on the veneer of their Wild West image.

Our trails crossed with the Padlock during the summer months when cattle grazed the high country. Our Forest Service grazing allotments above Granite Pass bordered their range, and often by summer's end a few cattle were mixed. We ran cows and calves and they ran yearling steers; thus it was usually Padlock cattle that came our way. Yearlings—the bovine teenagers—like to travel. They're curious, eager for adventure, and they have no baby calves to tie them down. In the fall when we rounded up, we'd swap cattle to put them straight.

"Look for deep tracks and thin shit," Mike explained. "That'll be Padlock cattle. Cattle handled wild get wilder, and those boys do run 'em wild!" Sure enough, when we gathered in the mountains, we heard tree branches crashing ahead of us and cattle bellering. We found tracks from cattle traveling at a run through the timber, hooves digging into the trail. When we saw loose, fresh manure, scattered thin, we gave chase—riding on the fly—lots of fun compared to our usual work plodding after cow-calf pairs. The Padlock steers, tails high, bucked down the trails and appeared at the creek looking like a bunch of bad boys caught in mischief. We'd put them with our herd until the Padlock cowhands came to get them.

In our yearly routine at weaning time, we hauled only the calves to the ranch headquarters, leaving the cows to spread onto better grass and put on extra weight without the nursing calves underfoot. The cows typically bawl around the pens for a couple of days, but they soon relax and scatter again, or as Mike said, "enjoy themselves, party a little."

One year, Stan asked Mike and me to ride ahead to Granite Pass where portable pens and a loading chute were set up, ready for trucks that would arrive later. Stan and our partner, Steve, stayed behind at the ranch to deliver cattle to a buyer, and they planned to join us when they finished. Since we were short of help, our son Dan and a couple of his junior high school friends skipped a day of school to help us corral the cattle, then sort enough calves to fill the first trucks.

"Don't waste any time," Stan said. "It's supposed to storm." That created urgency for the day: loaded semitrucks could not travel down the steep Shell Canyon switchbacks on ice or snow. We dressed for cold, wet weather—bulky coats, boot overshoes, mittens or monkey gloves, and Scotch caps. We didn't look very western, more like Pillsbury doughboys.

Mike and the kids and I knew the country and the gathering job, so we weren't worried about getting it done—just concerned about meeting the trucks on schedule. We corralled the cattle horseback and began to work afoot in the muddy, small pens that Mike liked to remind us could be "collapsible" if the cattle began to push or shove. I stood at the gate letting cows out and shifting calves into a side pen as we'd done time after time, year after year. The boys, barely taller than the cows' backs, quietly waved cattle toward me. Cows will usually see a gate, see the way out, and walk right past a good gateman as he moves slightly right or left, a careful rhythm that lets things happen without ruckus. The calves hesitate and lose their opportunity for escape, and the cows are outside the corral before they realize the babies haven't followed.

Mike stayed mounted beyond the gate, her lariat in hand—if a calf got out by mistake she would rope it and drag it back with no fanfare. We were working smoothly when the Padlock Ranch pickup wheeled in with a big, rattling, gooseneck trailer, the well-known brand painted on its side. We'd found some Padlock steers the day before when we rode the high pasture on Copeman's Tomb, so Stan had called their headquarters to relay that message, and we were expecting them to show up sometime. We paid little attention when we saw them drive up, as our own work was cut out for us, getting a load of calves sorted for the trucks.

A tall fellow took saddled horses from their trailer and led them toward the corral as another man in full cowboy regalia stepped out and clambered over the fence, walking toward me as he announced, "We'll just go ahead and get the Padlock cattle— won't take long; we'll use the corral if you don't mind. We're in kind of a hurry." Startled, I said, "No, sorry, you'll have to wait. We've already started." I knew the cowboys had to find those wild yearlings and then sort them out and corral them with Lord-knows-what commotion, and I figured we could not spare the time. We'd be lucky if they didn't "collapse" the corral. Besides that, his attempt to interrupt us was simply bad manners. The big fellow in his fancy cowboy gear shook his head and suggested we just step aside, and he offered to go ahead and do that sorting for us. He made it plain: that would be faster than waiting for women and some kids to get it done. He might as well have said, "Bless your hearts for trying, though."

He stepped over to take my spot, placing his hand on the gate. I didn't move. "No, we're doing fine," I said. "You'll have to wait." I was polite, I think. We faced each other for a moment—five feet of me, six feet of him not counting his Paul Bond boots with the undersloped heels and his high-crowned Stetson hat. Dan and the other boys kept their places too, continuing to press cattle toward me. The man barked some orders toward the boys, but they didn't glance his way; they watched me and the cattle—cows walked out, calves stayed in. We kept working, ignoring the buckaroo. After a few moments, he climbed out over the fence to sit in the pickup with his buddy. He put his feet up on the dashboard and crossed his arms across his chest, pouting like he expected to wait all day.

We had the calves sorted and penned when the trucks pulled up to the chute, and Stan and Steve arrived just then, too. They loaded our calves in short order, and then Mr. Padlock Cowboy became Mr. Congenial, since now he had real men to deal with instead of this women-and-kids outfit. The cowboys mounted up to find their ten or fifteen yearlings, a wild affair with ropes swinging and a lot of galloping and hollering. They finally got the steers corralled and loaded; they slammed the trailer door shut and the truck rattled away, the men sending good ol' boy waves for Stan and Steve, but not a glance toward Mike and me and our little crew.

Mike and I rolled our eyes and laughed. "Well," I said. "You're right. So it goes: Deep tracks, thin shit." Stan and Steve teased us as we explained what happened, and the boys giggled, for they, too, had felt the scorn of the big-timers, and they were proud of themselves for their own job well done.

"Yep, we'll see 'em *next fall if a-tall*," Dan shouted. "Let's go!" The boys jumped back on their horses and galloped toward Stan's truck where the lunchbox and thermoses of hot chocolate and coffee waited. "Eeh-hah! Ride, cowboy, ride! *Deeeep Tracks!*"

Mike admired skill and expertise in handling cattle, and when the *National Geographic Magazine* sent a big-time photographer to the ranch, Mike was horrified that he completely overlooked the things she felt worthy of a photograph. "Jee-sus Christ," she said with disgust. "Stan and I had that stray Mormon steer headed and heeled and stretched out right across Trapper Creek, and that fool never got a single picture."

The worst insult Mike could find for a cowboy was to call him a farmer. Cattle going up the mountain from Shell Valley pass through a lane on a creek bottom, winding up an incline called the Beef Slide. The wire fences in the lane were saggy—a joke, really—and the farmer there appeared each time cattle went by, standing in the gaps and flapping his jacket at the cattle to keep them out of his irrigated fields. Mike shook her head, time after time. "Those *farmers*. Why doesn't he just fix the fence?"

One October we helped the neighbors wean their calves at the set of corrals in Shell Canyon, right next to the highway, and when one of their cowhands cinched his horse up too tight, it bucked him off right on the pavement. He wasn't injured, but he was definitely embarrassed. Mike told the story every so often, shaking her head with the justification "Well, you know: he *was* just a farmer."

At some point in our ranching days, Stan's brother, David, bought an airplane, a Piper Super Cub, and he started flying around to check on things. While we gathered cattle, horseback in the hotter-than-hell desert or the icy winter hills, he zoomed that airplane over us. He tipped the wings and buzzed here and there, and we could never figure out what he wanted us to do or see. Mike would squint up at the sky and say, "Hmph. Why doesn't he just get his ass in the saddle?" She told another airplane story about her cousin in Buffalo who dropped water jugs from his little plane to the riders beneath him out on the Powder River roundups. She said he used rinsed-out Clorox bottles and "that smell lasted forever. Warm water to drink on a hot day, smelling a little bit like Clorox? Awful."

Despite an occasional swear word, Mike didn't act rough or tough; she was a lady through and through. Her cowboy clothes included a worn Stetson, a long-sleeved shirt, and soft leather gloves to protect her skin. She'd ride all day and still make it to the local supper club dressed to the nines, good-humored and ready to dance. She had impeccable manners and was at ease in any social setting. When Frank was elected to the Wyoming State Legislature, Mike gracefully took a place there with her husband.

She stayed on the lookout for anyone who enjoyed horses as much as she did. Three college boys came to Shell one summer, believing they could shoe horses for money and ride broncs at the Cody Night Rodeo for fun. Mike befriended them, and she let them set up a tent to live in at Horse Creek, planning to cook over a campfire even though summertime temperatures there often reach over one hundred degrees. They completely charmed Mike by asking her advice about horses and riding. "She just fell for it," Frank hooted. "I haven't seen a pie baked in this house in twenty years, and now here's Mike trotting down to the corral to take those damned kids a rhubarb pie. Or bringing 'em up here for supper." She vouched for them as cowboys and brought them to the mountain once, just when we needed help the most. I'd gone to bed that July night at the White Creek cow camp, wondering how on earth we were going to get a too-big herd of cattle trailed eleven miles through tourist traffic, to the U.S. Forest Service permit. At about midnight, Mike's trailer rattled into the camp bringing the three cowboys, their horses, and their bedrolls, and she made sure they were good help on the following day.

Mike and I pushed our own kids into becoming true friends and good hands although they resisted it, as kids will do. Her own daughter, as a child, was not eager to ride horses or become a cowhand; her sons were a little older than my children, and they didn't have much in common. Reluctant at first, they eventually met our expectations: they are now capable, dedicated cowboys and truly loyal friends to one another through personal crises of every kind. Mike's family still owns the ranch on Horse Creek, and they place the Rafter F brand on their calves each spring.

I knew she was older than me, but until Mike died at age seventy, it never really occurred to me that twenty-odd years separated us. In my memories and during our time together, it seemed we were the same age. Her heart suddenly failed, that September of 1989. Cautioning her about high blood pressure and some heart issues, her doctor had suggested that she exercise—walk, perhaps. Mike was shocked: "Walk? Me? On the Horse Creek Road? Hell, people will think I forgot where I tied my horse!"

I can see the "W" at Horse Creek from my window, and I think of Mike each time I glance that way. When the phone rings on a cold and snowy morning like this one, I half expect to hear Mike's voice: "What's going on down there? Want to run away? Want to come up for coffee?"

I smile and imagine that Saint Peter offered her a riding job that was just too good to turn down, and she took her string of top horses and rode up to the pearly gates where she was hired on steady.

11

Black and White

On this October morning, elk graze along the ridge behind our mountain cow camp. The majestic bulls make a spectacular sight with their huge racks showing boldly against the blue sky; below them, their harems wander across the sage-covered hillside. I'm sitting outside the old log bunkhouse—elevation eight thousand feet in the Big Horn Mountains—just passing the time. I reach for the binoculars to better see the elk as they spread across the slope. "Magnificent. No other way to put it," I say to myself. This is a small bunch—fifty or sixty—but last winter after the snow pushed in, we saw more than a thousand head in one herd.

Thrilling though they are to see, the big free-roaming herds are troubling to me and other landowners. The elk compete with livestock for feed and habitat, and their heavy traffic destroys fences that will need repair, come spring. Transfer of disease to livestock worries ranchers, and man-made conflicts around licenses, property damage, and trespass violations predictably occur during hunting season.

Unconcerned, these elk sun themselves on the brow of the hill, and I relax in the sunshine myself, enjoying a cup of coffee while I admire them. Suddenly, though, some squawking magpies appear beside the doorstep, distracting me with their raucous screech and chatter. They flap their wings, diving boldly at each other. Although splashy in black and white, they seem gaudy and annoying on this peaceful morning. A phrase pops into my head: "It's not always black and white." Smiling, I can almost hear Governor Mike Sullivan's voice echoing back to me from years gone by.

Governor Sullivan called me in 1991 to ask if I would accept a six-year appointment to the Wyoming Game and Fish Commission. "I have to warn you, it won't be simple," he said. "There's a lot in the works right now—federal programs, wolves, grizzlies, bison, private property acquisitions, access, and other issues that are really heating up. You'll have to be approved by the Senate. Can you do it? There are two sides—at least—to everything. Things aren't black and white."

Our ranch was enjoying a bubble of optimism and prosperity just then; our children were grown, and it seemed the timing was right for me to make this contribution of service. I said I thought I could handle the job and naively stepped into an uproar I could not have fully imagined. Other ranchers, and, in fact, other women, had served on the Wyoming Game and Fish Commission, but in my case things quickly got personal. "Ag Figure Named to Game and Fish," the *Billings Gazette* headlined. A letter to the editor of the *Casper Star-Tribune* begged voters to "call your senators today . . . and urge them to vote against confirmation of Mary Flitner." I didn't think of myself as an "ag figure," just a citizen willing to serve the State of Wyoming. My credentials seemed mundane enough: Library Board, School Board, 4-H leader, mother, housewife—and yes, rancher. I was stunned by the outrage directed toward me from people I had never met.

The press ranted about my "connections." My husband, Stan, had good standing among range management and conservation policy makers, and served as an officer of the Wyoming Stock Growers Association. And the newspaper named other names: my brother-in-law, president of the Wyoming Farm Bureau; my cousin, executive director of the Wyoming Stock Growers Association; a different cousin, an attorney with expertise in private-lands issues; another cousin, a woman and a Democrat, serving in the Wyoming State Senate. I had numerous friends in the sheep industry, too. My family and Stan's had a long history of public service and involvement in natural resources venues, and it seemed to me these connections would provide insight and expertise if I had questions to ask.

I had friends—connections, it could be said—in the Nature Conservancy and other environmental organizations, too, but those weren't mentioned in the newspapers. "In a conflict between agricultural interests or sportsmen's interests, which side would Mrs. Flitner support?" Speaking to the *Billings Gazette,* a spokesperson for a sportsmen's advocacy group called my pending appointment "an insult and a slap in the face to hunters and fishermen."

With some confusion, I responded to reporters, saying I had "no crusades going, one way or another. Harmony has been unnecessarily missing between ag and the Game and Fish Department. We need to be talking with each other rather than against each other."

The uproar continued, though, and I decided to drive the four hundred miles to Cheyenne for the Senate's vote on my appointment, to appear in person and face the opposition. Aiming for a conservative, dignified appearance, I carefully chose a business-like wool dress and my best high-heeled shoes, which I hoped would show me as professional, sensible, and nonthreatening. I parked the car at the Wyoming State Capitol Building and nervously climbed the long bank of stair steps that approaches the front door.

I took a deep breath and walked inside, surprised that the lobby buzzed with laughter and conversation. Someone greeted me, and someone else stepped over to shake my hand. I discovered I was surrounded by people wearing badges that said "She's *my* cousin, too!" Yes, some of the people *were* my cousins—Republicans *and* Democrats. Most, though, from Cheyenne and other parts of Wyoming, were not my relatives at all. Some were only participating for the sheer fun of joining the fray; others had come to show resistance to the unfairness and inaccuracy of the press releases and letters to the editor; still others were there to support my nomination. Having so many "cousins" didn't have anything to do with anything, but I was glad to have the boost to my confidence.

I watched from the balcony of the Senate Chamber as the votes were cast, yea or nay, to confirm my appointment. When the Senate leader called for the vote, I remember that Fremont County senator John Vinich, a Democrat and *not* a rancher, *not* my cousin—a man I barely knew—stood on the Senate Floor to say, "In Wyoming, we don't approve or disapprove of anyone because of their relatives. I vote to approve." The vote confirmed my appointment.

"Two sides—at least—to every conflict," I commented to the newspaper reporter, "and I believe I can evaluate them fairly and honestly."

Years have passed since I finished my term on the commission, and on this October morning, the magpies have nudged my memory of a different, funnier story about wildlife, and black and white, and conflict.

When I was a child in tough-winter, deep-snow Sublette County, my grandpa and my dad, like other ranchers of their time, used a draft-horse team and sled to put out hay for their cattle. Dressed against the God-awful cold, they harnessed and hitched the horses, then drove them out into the meadows where the cattle waited beside a fenced-in haystack. Stacks were loose hay then, the pretty dome-shaped mounds seen in paintings and old photographs. The men forked hay out of the stack onto the big four-runner sled and then pitched it off to the cows, one fork at a time—exhausting work—as the horses pulled the heavy load. The ranchers used the entire summer to irrigate, harvest, and stack that hay, one pitchfork-full at a time and, paradoxically, spent the winter pitching it back to the cows.

Wyoming law provides that management of wild game is the responsibility of the state, specifically the Wyoming Game and Fish Department. Here's the rub: the State of Wyoming controls the issuance of hunting licenses but owns little of the necessary habitat. Federal agencies—Bureau of Land Management or United States Forest Service—provide some forage for wildlife, but Wyoming is roughly half-and-half, private lands and public lands. Private lands generally offer good water and feed, including hay and other crops, nowadays. "Those old-timers weren't dumb," I've heard more than once. "They filed homesteads on the best land and left the rest unclaimed."

Elk, deer, antelope, moose, and other wild animals don't observe the property lines, however. It's not "black and white"—now or in Dad's day. He and his neighbors tried not to begrudge range grass for deer or elk, but the stored, stacked hay was precious, hoarded against cold, brutal winters. For a rancher and his livestock, that hay could determine survival or not.

Dad liked to tell of such a winter in the 1940s or 1950s when each day a large herd of elk persisted in climbing down from the ridges onto Dad's meadows, crowding the cattle for hay at the feed grounds. On a below-zero morning, Dad telephoned my uncle Francis Tanner in a panic, saying breathlessly, "*Tanner,* I need some help. *Now.* You've got to come *right now.*" Tanner was Dad's best friend and his brother-in-law, a lively, fun-loving guy who ran the International Harvester dealership in town.

"What?" Tanner asked. "Why? What's going on?" Dad knew the party line "rubber-neckers" might be listening, so he said, "Never *mind.* Just come, right now. Right *now.*" Well, Tanner came, and Dad showed him what: at the feedground that morning Dad saw that elk had broken into the stackyard, stomping the haystack into ugly, dirty heaps and leaving the fence sagging, wires broken. The cattle restlessly circled the stackyard, confused and hungry, while the elk sunned themselves, bellies full. Some lay on the stack, while others lounged across the feedground.

"All my troubles showed up right there, that minute," Dad later said. "The cattle prices, ranch debt, no hired help, not enough hay, bills to pay, the awful cold, the snow two-feet deep—and then the damned elk—it was more than I could stand." Dad carried a rifle with him; he fired a few shots to scare the elk away and then emptied the gun in earnest. Said and done, several elk lay dead and the others ran off. "For a few minutes," Dad said, "I just felt enormous satisfaction at taking care of those sonsabitches, but then I realized I could be in big trouble." That's when he called Tanner.

Dad and Tanner fed the cows and then loaded the now-frozen elk onto the hay sled. They hauled them back to the ranch to hang in a shed, thinking they'd make use of the meat. Tanner's wife, my aunt Helen, brought their three boys out from town for supper, and we cousins giggled around the edge of the grown-ups' rowdy celebration.

Food and a little whiskey were getting the evening's hilarity under way, when someone knocked at the door and Dad answered, surprised that anyone would show up just at suppertime. There in the porchlight stood Game Warden Boyd Charter, Mr. Local Law Enforcement himself. "You're under arrest," he said to Dad. "I *know* you shot some elk today, out of season, and you harassed the others; that's against the law, too. I am here to confiscate the carcasses."

"You don't know any damned such thing," Dad said. "And you won't set foot on this place without a warrant."

Dad and other ranchers had complained about the large numbers of elk that year, demanding that the Game and Fish Department find a way to drive the elk away, feed them somewhere, reduce the herd size—do *something*—and Charter had hired a plane that particular afternoon to fly over the Meadow Canyon, finding facts for himself. From the air he saw the snow-covered meadow, the stackyard, tracks, blood in the snow, hay trampled and ruined, and the cattle nearby. No elk in sight, but plenty of tracks where they'd gone back up the ridge. He saw sled tracks leading back to the ranch buildings, and he guessed what had happened. Charter had no tolerance for "big-shot ranchers" as he called them, and he often bragged that he did not abide poaching or violations.

"You're under arrest," he repeated.

"Get the hell off my place," Dad told Charter. "You can't prove anything." Their loud exchange of insults and swearing carried to the top of the stairs where we cousins huddled, listening, wondering if our dads were going to *jail*. Charter finally left saying, "I'll get that warrant, and I'll be back. Count on it. You won't get away with this."

The minute Warden Charter's taillights disappeared, Dad and Tanner ran to load the elk carcasses onto Dad's old truck. They had to get them off the place; they had to think of a way to dispose of the evidence. "We figured Charter wouldn't be gone long and we had to do something, fast," Dad recalled.

They roared the truck through ruts of snow on the county road, not sure what to do in the dark of night. A few miles down the road, they crossed the North Piney Creek Bridge, and he and Tanner had an inspiration. They stopped on the bridge—bare of snow—and they heaved the elk carcasses under the bridge, out of sight. They didn't even veer off the road. No tire tracks, no footprints, and no blood, because by now the carcasses were frozen solid. Dad and Tanner hurried back to the ranch to wait. When Charter returned with the warrant, he looked through the sheds and the barn, but he found no trace of dead elk; without evidence he couldn't verify who'd done the deed, or even if a deed had been done.

"He never did figure it out," Dad would say with a big laugh and a slap on his leg. "He couldn't prove a thing. Charter never knew what happened to those elk. We never told anybody anything. The thing was, though, magpies flew out from under that bridge all winter long. Every time a car crossed, made that *ka-whump* sound—boy, those magpies flew! Black and white, everywhere. *Dozens* of 'em! They feasted for months. We were scared to death somebody would find those carcasses, but nobody ever did."

Decades later, in the 1990s, my fellow Game and Fish commissioners and I faced the same conflicts, but without the laughs. Animosity over damages to crops and the Game and Fish Department's seeming disregard for private lands continued, and controversies were stirred by sportsmen who, by then, unlike in Dad's day, had big 4-wheel-drive pickups and ATVs. Fees charged for hunting access onto private lands were becoming common, and sportsmen wanted more licenses, more access, and a bigger game population. Wildlife advocacy groups pushed a slogan, "Public Lands—Cattle-free by '93!," suggesting that wild game would stay on National Forest or Bureau of Land Management property if it weren't for the grazing of domestic livestock.

Previous commissions had acquiesced to Game and Fish pronouncements, but I felt that honesty and direct discussion would come closer to solving the recurring problems. The *Casper Star-Tribune* continued to profile me as a rancher with bias against the needs of Wyoming's wildlife population. I tried to speak out honestly, even when my views weren't popular. I felt that private landowners should be recognized for stewardship of the land in general and commended for the private-lands habitat and feed they provided to game animals. During the discussion at a particular meeting, several of the other commissioners agreed when I stated my opinion that the Wyoming Game and Fish Commission should publicly acknowledge that "half of the State, and the habitat therein . . . is privately owned." On August 30, 1995, a *Star-Tribune* headline decried that our commission "excoriated" the Game and Fish Department with a "sharp public tongue-lashing" in reference to the staff's disregard for private lands. The *Tribune* also said we'd "castigated the department for failing to recognize the contribution private landowners make to healthy wildlife populations."

Perhaps we had. I liked and respected most of the staff members, and I worried when I saw that headline—perhaps I had overstepped. I phoned the Game and Fish director saying, "John, do you feel *excoriated?* I didn't mean it that way—it was just an honest opinion." He laughed. "Well, it got my attention," he replied. I was relieved that he hadn't lost his sense of humor.

Cattlemen, sheep producers, and farmers didn't approve of everything I said or did, either, and they weren't afraid to seek me out or to demand answers, to express their dissatisfaction in person, or to bring complaints to me—but without holding a grudge when decisions didn't go their way. The agricultural industry straightforwardly, sometimes noisily, scrutinized commission positions about damage claims, license policies, and, later, the "experimental, non-essential re-introduction of the wolf," which our commission reluctantly approved, and which has not been satisfactorily resolved as of present time.

Magpie-like chatter continued to resound from my critics, but I never regretted that I spoke out, or that I asked questions, expecting answers. Conversations eventually did develop between conservation groups, wildlife agencies, environmental advocates, and private landowners, bringing understanding and civility amid opposing views. Our commission of the 1990s opened some of those doors, possibly because of the publicity that followed my appointment.

With the passing of time, from my father's era to the present, Wyoming's concerns for wildlife have spread far beyond the simplicity of hay in a haystack. The State of Wyoming Game and Fish Department now addresses topics such as endangered species, birds, amphibians, fish, wolves, and grizzlies, and the effects of gas and oil exploration, drilling, and open-pit mining. Rural sprawl and subdivision and housing developments seriously threaten migration corridors and the general well-being of wild game animals. In some communities, deer lounge on the lawns in broad daylight, as though they are living lawn ornaments. Ranchers, sportsmen, conservationists, and Game and Fish personnel sometimes find themselves in agreement—a positive change.

Some thoughts linger as I consider my time as a Game and Fish commissioner. Early in my term, I sent a memo that stated my determination to "make myself heard, no matter how hard that sometimes is." In the shoebox of newspaper clippings and old Game and Fish correspondence I've saved, I recently ran across a card dated March 1991, shortly after I was appointed to the commission. A young woman wrote, "I've been following with interest your appointment to the Game and Fish Commission. It is a well-deserved honor. You are an inspiration to the women of Wyoming." Women had previously served on the seven-member commission, and, in fact, one woman served with me. I hope the card referred to my determination to speak out frankly and openly, without regard to gender or stature. I eventually did receive some positive recognition from agencies and organizations, but nothing pleased me as much as this card.

I worked comfortably with most men and women I met, and I had no ego to protect. I was glad to listen. Discussions didn't always lead to agreement, but often, at least, they led to understanding and mutual respect. At the end of my term, I left behind a memo that said, "I believe our Commission changed the Department perspective from 'nothing is possible' to 'some things are possible and, in fact, inevitable.'"

That's a lot to think about on an October morning, though. The sun pushes gold to the hillsides and the elk meander along, nonchalant, unworried. The magpies continue their raucous bickering, disturbing the morning's quiet and halting my reminiscences. A sharp breeze begins, too, so I stand to stretch and pour out my now-cold coffee. I think of the approaching winter and I smile, thinking especially of my dad. I can hear his voice: "Boy, they flew! Black and white everywhere! Magpies everywhere. Boy, those sonsabitches flew—just a cloud of black and white."

Budd family, 1950s, at Meadow Canyon Ranch, Big Piney, in the Green River Valley of southwest Wyoming. *Front row, left to right:* Mary Budd (author), cousins Dick Tanner and Bob Tanner, sister Nancy Budd. *Middle row, left to right:* aunt Helen Budd Tanner, grandparents John and Lula Budd, sister Betty Budd, mother Ruth Budd, uncle Francis Tanner. *Back row, left to right:* cousin John Tanner and father Joe Budd. *Photo by Record Stockman photographer Ross Miller.*

Author's grandparents, Lula McGinnis Budd and John Budd, circa 1950, near Big Piney. Grandma is on her mare Calico and Grandpa is on his horse Joker.

Budd cowgirls at Meadow Canyon Ranch, probably 1947. Betty is on Snip, Mary on Brown Jug, and Mom (Ruth Budd), holding Nancy, is on Smokey. The original ranch barn and corrals are visible in the background.

Author's father, Joe Budd, and his father, John Budd, partners at Meadow Canyon Ranch in the 1950s.

This early 1960s livestock magazine ad, showcasing a scene near Big Piney, Wyoming, reveals changing times for the Budd family. Meadow Canyon Ranch became Budd Ranches, under the ownership of Joe Budd.

The Flitner family, 1950s, at Diamond Tail Ranch, in the Big Horn Basin of northwest Wyoming, near the town of Shell. *Left to right:* John Flitner, Patricia Flitner, parents Howard and Maureen, Stan Flitner (author's future husband), and David Flitner.

The Flitners' "Cowhide Sign" near Shell, Wyoming, a local landmark for decades.

Stan Flitner and a little cowhand, along with family cowboys, driving cattle through the Mail Trail, 1973. The trail is headed out of the desert toward the high country.

Flitner family ranch partners, 1978. *Left to right:* Stan and Mary, Howard and Maureen, and Sue and David. The Diamond Tail ranch barn is in the background.

Stan Flitner harnessing the workhorses, 1980s. *Photo by Bob Budd.*

Mary Flitner, horseback at calving season, 1980s.

Mary with the branding iron, spring 2015. *Photo by Sherre Wilson-Liljegren.*

Mary with friend "Mike" Tisdale Hinckley, horseback in the Big Horns, summer 1985.

Cattle drive toward Granite Creek. *Photo by Lee Raine, www.cowboyshowcase.com.*

Mary and Stan with cattle at Snowshoe Pass. *Photo by Lee Raine, www.cowboyshowcase.com.*

Mary at cow camp with mare Savvy. Note the Diamond Tail brand overlaid on the bridle's silver conchos. *Photo by Lori Hinman Dorr.*

Jamie Flitner, Anna Flitner, Jane Bell, and Carolyn Walton—"all girls"—pushing cattle down the trail. *Photo by Lee Raine, www.cowboyshowcase.com.*

Mary and Stan with saddle horses at the Mackey cabin during late autumn.

Cow Camp headquarters on White Creek, looking toward Snowshoe Pass.

1910 photo of the "W."

The "W" in wintertime, present day.

Cattle heading home to Shell Valley, coming off the Beef Slide.

Mary and Stan Flitner with all the family at Diamond Tail Ranch centennial reunion, 2006.

12

The Castle Guardians

December 2005:

We sold Castle Garden Ranch in the fall, a big event.
I know it was the smart thing to do—we doubled our
money and the partnership had started to fray as
everyone got so busy with other obligations. After all,
the ranch did what it was supposed to—gave each of us
a new foothold. I have never sold any land before, and
I don't like the feeling.

We were part owners of this ranch in the very center of Wyoming
for only one decade. I squirmed a little to think that we had barely
passed through, used what we could, and moved on.

Ten years before, in July of 1995, our friend Dennis Horton had
called. "I'm looking at a ranch that's for sale, near Moneta. My
brother Darrell is going to partner with me, but we can't put it
together alone. It needs about a thousand pairs to make it work,
and we don't have enough cattle to stock it. Are you interested?"

"A chance of a lifetime, a hell of a good deal," Dennis added.
"Six months of grass, mostly BLM, some deeded land. Come and
look," he said. "It'll pencil. Price is right. Think about it."

"I don't have to think about it. No, thanks," I told Stan. "It's a
hundred miles from home, and we're already spread too thin."

Just a few weeks earlier I'd written in my journal:

> Rock bottom. Dan and Tim and Stan and I have been
> going as hard as we can and no two of us can agree about
> anything. Tim is skin and bones. Every day I am tired,
> tired. The whole summer has been one disagreement after
> another if anyone slows down long enough to talk. Not
> a minute without pressure—from work, from the bank
> and the cattle market, from two sons reaching for respect
> and responsibility. From the weather, from being short-
> handed, from machinery breakdowns, calamities with
> livestock, horses, wire-cuts, miscommunications—all
> rubbing like grains of sand in a shoe.

"No," I repeated. "We just can't."

But later I wrote, "I guess we're going to do it." Stan, Tim, and Dan looked at the ranch at Moneta with Dennis, and they agreed: it would pencil. We could expand our cattle herd to a more profitable size, and I could hope a new venture would bring my husband and our sons into a comfortable working relationship. Perhaps they could learn to lean on each other, instead of pushing against each other.

Dennis—a high-voltage farmer and a cattleman himself—fearlessly led us toward the investment. His brother Darrell—amiable, honest, and steady—came on board with his son. Another family joined us—Mike and Beth Evans—making four sets of partners. Mike and Beth were smart and capable, fun-loving, leaving good jobs to go "on their own." We did not know one another well—only that each partner had reason to grasp at opportunity.

We met with banks, attorneys, and accountants, and then there we were: owners of the newly named Castle Garden Ranch—in the geographical center of Wyoming—almost 100,000 acres of sagebrush, shallow draws, and sandy benches, fenced in the shape of a monstrous cowboy boot.

The following spring, 1996, Stan and I caravanned to Moneta to prepare for the first season with cattle at the ranch. He drove a big truck hauling material for reservoir repairs, and I followed with the pickup and horse trailer. "I had a flat tire on the way," I wrote in my diary. "I was glad not to have to fix it alone, as my rig was fully loaded."

We arrived at the middle-of-nowhere headquarters where we unpacked our sleeping bags and a few groceries for supper. We assessed the rickety set of corrals, the shabby house with the bare necessities of electricity, plumbing, and sulfur-smelling well water—no telephone, no nearby town. "I can't help feeling kind of excited," I wrote. "The grass looks good and there's plenty of livestock water, so far. Beautiful sunset, wide-open spaces, swallows and larks all about. Antelope and a coyote, too."

I shivered at the loveliness of soft gray and blue sagebrush, straw-gold grass, flame-red sunset, drifting clouds, and vast reaches of rangeland. I saw miles and miles of sandy draws and gentle hill-slopes—nothing like the brutal canyons, rocky outcrops, and steep rims circling the fenced pastures and farms in the Big Horn Basin. Here at Castle Garden Ranch, no rushing creeks, no rivers, few trees, few rocks, and just the one arterial gravel road and a confusing web of old ruts, with Muskrat Creek dribbling through the middle.

Our new ranch encircled the historic Love Place, where early stockman John Love homesteaded in the late 1800s. At its peak, Love's ranch, described by John McPhee in his book *Rising from the Plains,* ran eleven thousand sheep and several hundred cattle before it fell—defeated by weather and markets—a testimony to the risks we might face here, as well as the land's potential for livestock grazing.

A color-coded map showed the private land tracts we purchased and the predominant, enormous areas of public lands administered by the Department of Interior. The BLM agent made himself clear: "This area is complicated with public lands issues—ancient archaeology sites, wildlife, oil and gas development, recreational use, *and* domestic livestock. We aren't interested in new grazing schemes or cross-fencing or any other ideas. Just do your job, we'll do ours, and we'll get along." Only one parcel of deeded land adjacent to the headquarters was fenced separately; the remaining acreage had only the perimeter boundary fence. Thus we were bound to operate according to BLM directive, paying the government for the days of use as is the requirement.

"Six months use. That's it," the BLM man went on. "You can operate the windmills. There's a little live water, some stock ponds if the weather gives you help. But there's plenty of grass for the numbers we've allocated to you. We save the other six months' grass for wildlife."

"It's turn-key, once we get the place in shape," Dennis said. Lanky and fair-haired, always in a hurry, he made it sound easy. "We unload the cattle in May, keep an eye on the waterholes. Then we each take our cows home in the fall." Oddly, until they got acquainted, the cattle actually grazed in separate herds—they weren't comfortable with each other, any more than we partners were, at first. Eventually the cattle blended into a practical, working unit, and so did we.

The Hortons liked to say they were farmers, not cowboys—but they ably provided machinery and knowledge to repair windmills, reservoirs, waterlines, and tanks. Stan, Tim, Dan, and I offered experience with handling livestock and gathering big country, and we organized that job with Mike and Beth—once each September to wean the calves and again each November when we hauled the cows home. None of the four families moved there permanently; we were in and out each year for intense days of riding and sorting cattle or for a major repair job, always driven by the calendar and the weather.

A wobbly picket fence circled the house at headquarters. We partners collected thrift-shop dishes, a refrigerator, cookstove, table and chairs; an old TV worked if the tinfoil was placed just right on the antenna, and we found cots for the three bedrooms. The garage held tools, fence supplies, gasoline cans, ropes, sacks of concrete, oats, and pack rats. Deep sand had discouraged somebody's try at a lawn, but the three little Evans boys—Chad, Mark, and Jess—played for hours, excavating a make-believe road system with their yellow Tonka trucks and tractors.

Oddly, facing problems in a new setting seemed a relief. At Castle Garden, the tasks were obvious, without drama: scrub, clean, wash, sweep. Ride, learn the terrain. Fix the fence. Repair corrals and water tanks. Leave the confusions at home, at home.

In the early years we didn't know the country, and we could not expect that the cattle would find trails or follow them as they eventually learned to do. At first we encouraged visits from wives, girlfriends, boyfriends, children, friends, would-be cowboys, photographers, and visitors who wanted to ride, as though it were a holiday. In that huge expanse of land we thought a big crew would be useful. We learned that all those riders meant too many horses for the corral, too many people who needed beds and meals, too many lunches to pack. It meant too many horse-trailers and cowboys to transport, too many people to worry about—to direct and keep safe. We finally avoided that practice and used only cowboys with credentials—the partners, mostly, and what good, able help we could find.

"Sometimes, less help is more help," I told Beth when she asked how many potatoes we'd need for tonight's stew. "One apiece and one for the pot," I guessed. "That makes twenty-three today. And the same for breakfast." We planned the menus at home before each roundup. I still have our work lists:

> Bring from home:
>   25# hamburger
>   beef for roasts
>   30-cup coffee pot

thermoses
electric pancake griddle
big cook-pot
big electric roaster
2 big skillets

Buy on the way:
pancake flour, syrup
lunchmeat
ham
bacon
eggs
canned vegetables
apples and oranges
bread
ketchup, mayo, mustard
coffee, juice, milk
potatoes, onions
chickens, and noodles
ice, dish soap
candy bars
scotch for the cooks, everybody else
bring their own drinks.

In the early years, the lists included "juice boxes" for the little Evans boys to carry horseback. Young as they were, they could ride all day if they just didn't run out of juice!

Each morning I met Beth in the kitchen. We sleepily poured mugs of coffee from the big, automatically timed coffee urn—whispering, sharing laughs, rolling our eyes or shaking our heads in that quiet, still-dark interval before the action started.

We cooked bacon, eggs, and pancakes for breakfast, then packed thermoses of coffee and coolers of sandwiches and water while the men saddled and loaded horses. The trucks chugged out at daylight on two-track roads; sometimes Dennis would scout ahead with his motorcycle to spot cattle in remote places. Frequently our crews met during the day—at Boystown on the south, or at the main corrals—other times we didn't cross paths at all, reuniting at headquarters in the evening. Our work was a far cry from what John Love and other early settlers endured but, nonetheless, a big job. I had trouble learning the landmarks in such vast country with no mountains to give me bearings, and I was glad we worked in teams, keeping track of each other and the cattle we found.

Beth and I rode long days with the men; we liked it, we were good at it, and they appreciated us. Her pretty red hair, quick smile, and practical Greek-Irish work ethic made her a jolly soulmate. Each night, back at headquarters and after the big supper was finished, the men did the dishes while everyone relaxed—Beth and I each with a glass of scotch and the men with their own drinks. Stories from the day got funnier.

"That guy, he rides like he's coming apart!" Duane said about one of the cowboys.

"I couldn't find any place to put Mal, where he wouldn't be in the way," Dan said, "but he was so anxious to *help* when those cows knocked the gate open! So I told him, 'Mal, go let the dogs out!'"

"In the whiteout," I said, "Stan and I tried to follow Muskrat Creek out to the road. When we met an old lame cow plodding through the snow, I said, 'That's the crippled cow we dropped *behind,* to come in on her own.' Stan got off his horse and broke the ice in Muskrat, to see which way the water was flowing. We were lost! We'd made a circle and were going the wrong way!"

One night at camp, cold and wet from a day like that one, someone noticed that our friend Larry had kind of faded, tilted a bit in his chair after a shot or two of whiskey. "Larry, you doing okay?" He raised himself up with a satisfied smile: "I used my compass today to find headquarters. And I am just *pickled tink,*" he said.

During one roundup we ran out of sandwich bread, so Beth and I left urgent messages for Darrell, who was trucking cattle that day. "Get bread." We sent word to his house, his business, the gas station, and anywhere we thought he might stop. When we reached the bunkhouse that night, a pile of bread loaves sat on the table beside a note: "To Beth and Mary, head honchos—this could turn out to be a helluva day. I got bread at Greybull, bread at Basin, bread at Worland, and with any luck I could get *bred* when I get home tonight."

Some evenings we crowded around the table to plan the next day's work. We argued, interrupted, and corrected one another as we ate our meals or finished a drink, juggling the unknowns of personalities, geographies, livestock, and weather. Time was always short, sometimes tempers, too; frustration only a word away from real anger.

One night we reached flashpoint. "I've got to get home. I've got work there to do," Dennis said. "Everybody does," Stan said. "You aren't the only one. But nobody ought to leave till we find all the cattle." A strained, awkward quiet followed; hard looks crossed the room. Suddenly Mike—shirt-tail out, a three-day beard, saggy Levi's—jumped from his chair, smiling broadly and waving a yellow tablet. "I've got it!" he yelled. "A No-Huddle Offense!" He grabbed a pen, scribbled stars and lines and arrows, rattled names and orders and directions. "Tomorrow! Daylight! Saddle Up and *Go!*" Mike slapped the tablet onto the table. We sat dumbfounded for a minute or two, then whooped into laughs, sidestepping the tension. Too many quarterbacks, for sure. All chiefs, no Indians. All bosses, Type-A personalities. All leaders, no followers.

Tricky business, this teamwork, but somehow we made it work as we learned to know and trust each other's skills. On the evening after Stan's horse threw him and he broke some ribs, he went back to Shell early. The rest of us sat at Castle Garden with our drinks, considering options. Mike said petulantly, "How will we get along without *Stan?* He's our *scapegoat!*"

Those ten Castle Garden years sit apart in my memories, as few experiences stand so distinctly, identifiably. The weather determined success or failure for each excursion, and the timeframe was critical: get there, get the work done, get home. Naturally, the cattle reacted to the weather, seeking shelter or feed, and temperatures reached every extreme: hot, cold, wet, dry, snow, ice. We cussed the relentless wind that blew grit and dirt, filling our eyes, blasting our skin. We brought overshoes, slickers, warm coats, light jackets, hats, ear-flap caps, coveralls—it was anybody's guess which we'd need, and over the ten years I believe we used every combination.

The landscape, beautiful and huge, did not prove to be as threatening in its size as I had feared. The flat, sandy soil was easy to traverse horseback, and once the cattle learned to trail to water, we only had to follow the trails. If I saw tiny dots in the grey sagebrush miles away, I knew that they were cows, and that I could trot my horse there without his breaking a sweat, conditioned as our horses were to rocks and steep pitches at home. I could bring those same cows back, and every time see a new landmark—a gully I had not noticed, a sheepherder's monument, or perhaps an abandoned cabin site.

Our personal trails took new twists, too. At the beginning, Stan and I were in our fifties, while our sons Tim and Dan were young men, savvy and aggressive. Stan and I wanted their participation in our family ranch, but we had not yet found a way to accommodate the necessary changes. Those ten years from 1995 to 2005 held uncertainties as we struggled to understand one another. Tim and Dan both got married. Tim and his wife had a baby girl; Tim was diagnosed with cancer; then he and his wife got divorced. He endured chemotherapy twice, the second time while Castle Garden Ranch was being sold.

In that same ten-year pocket, Stan had heart surgery within a few weeks of breaking his ribs, leaving Dan to fill in the gaps of a hunting business and work at home while Tim and I tried to cover the responsibilities at Castle Garden. Our daughters, Carol and Sara, gave us strength and encouragement, and we tried to reciprocate as they built their own lives, families, and careers in Wyoming.

The world kept spinning, of course. I finished my six-year term on the Wyoming Game and Fish Commission. Stan served as president of the Wyoming Stock Growers Association. In November of 2000, we were at Castle Garden during the Bush-Gore presidential election, and we watched the snowy-screen television, listening and laughing with the rowdy cowboys as the results came in. Stan and I went to bed in the bunkroom that night surrounded by the not-so-little-anymore Evans boys in sleeping bags, believing Al Gore had defeated George Bush. We weaned calves at Castle Garden just following the 9–11 World Trade Center bombing in New York City, and we talked of nothing else that fall of 2001.

By the time we sold Castle Garden Ranch, Dan, his wife, and baby boy took advantage of an opportunity to ranch in New Mexico, and Tim remarried. Stan and I built a new house, and Tim and his family moved into the old ranch home back in Shell Valley.

The livestock market roller-coastered high and low throughout the decade, and at the end of the ride, we sold calves at a record-high price. We sold cow-calf pairs, too, and seized a chance to sell timber from our deeded mountain land. With the sale of Castle Garden Ranch, Stan and I and Tim and Dan were debt-free for the first time in our ranching lives.

Years earlier, our neighbors around Moneta might have wondered if we would fail or how we dared plunge into a country that had historically chased others away, but we progressed from self-conscious newcomers to becoming their friends. I could not blame them if they disapproved of our step-in, step-out seasonal presence there, or if they wished for ranchers who would stay and offer commitment to that region and community. I think, though, we left the ranch better than we found it. We did not erect a pretentious, eye-stopping gatepost or make a lot of changes as new owners often do. Instead we looked after the land, and then we stepped away with humility and gratitude and respect for past histories.

Castle Garden Ranch ended as it started, with Dennis finding the deal: "I talked to a guy who wants to buy the ranch. We can double our money. What do you think?" And we said reluctantly, "Yes, it's time."

That last roundup, we had perfect fall weather. Clear November mornings, sun-filled cloudless afternoons, blue-orange sunsets, and only a gentle, humming wind. We had a skeleton crew, unlike the times we'd stuffed too many people into the little bunkhouse with muddy boots and wet coats, with dogs fighting for a place to rest in the garage, and wall-to-wall sleeping bags.

We were short of help, the proof of "time to sell." It is a cowboy's ranch, requiring attention and commitment if only for those few times a year.

This time Stan and I took the outside circles, although we were the oldest riders on the crew—in our sixties by then. We had great horses to ride that year, each other to count on. We rode the circumference of each day's country, gathering cattle and pushing them toward the big herd that trailed to the loading corrals. From miles away I saw cattle coming down the now-familiar draws, long thick walking ropes of cows, or straggly little bunches, here and there.

We knew we wouldn't be back to those special spots: the huge sandstone turrets and petroglyphs at Castle Garden, Fish Draw, the big rocks I called dinosaur eggs, the Moneta Divide, the Love Place, Corral Gulch, Mahoney Draw, Jumping Off, Muskrat

Creek, Squaw Butte, Chalk Hills, Boystown, Fraser Draw, and the skylined monuments placed by somebody decades or centuries ago. I tipped my hat a time or two—a comic good-bye.

In the end, we partners met at the bank to close the deal in our customary style: a hurried lunchtime hamburger instead of the drinks-and-toasts ceremony Beth and I wanted. We parted with a wave. "Work to do at home. See you later."

I hated closing the gate on Castle Garden Ranch, taking a last look around, seeing no cattle, and knowing I would never see ours there again. When I drive between Shoshoni and Casper now, years later, I nod at the road turning south. "Castle Garden," the sign says. The wind boosts the tips of the sagebrush and bends the grass, and I smile. Yes, we were the Castle Guardians. It was a privilege.

13

May I Have This Dance?

"A perfect gather is like a dance," my dad said. "Everybody keeps in step, keeps the rhythm."

In my mind's eye I'm with Dad in the grey dawn. He is kneeling on the ground beside his horse, holding the bridle reins in one hand and scratching a map in the dirt with the other. Cowboys peer over his shoulder. We've trotted miles to pause together on a high sagebrush ridge where each rider learns which is "his" country to "clean." "It's all about timing," Dad says, his short-brimmed felt hat pushed back on his head. "Watch your partners. If you get too far ahead of other riders, you won't see the cattle coming toward you. If you're too slow, you won't meet." With a stick, he scratches lines from the outside of an arc into the center where he's made an X that he says is a reservoir. He tells the cowboys again: "Pay attention, work together. We'll meet in the middle."

Today, sixty-some years later, I find myself in the same spot for this big roundup day. My sister has asked me to help gather several hundred cattle in Sublette County, at the old home ranch. I haven't ridden here since I was a child, and I can hardly wait to dance in this country again. I shiver in the chilly morning, wondering, "Do I know the steps? Do I remember this country? Will I keep up?"

A good gather on any day, in any ranch country, is a lovely, rhythmic reel—a complicated line of dancers together bowing and spinning gracefully across the landscape. Horses' hoofbeats set a soft clip-clop tempo as we step into the ballroom of the day. "May I have this dance?" We nod to each other; we begin. I hear an occasional melody of birdcall and the crooning of cows to their baby calves. A bull in tenor, an answer in bass from a bull far away. *Forward, back, heel and toe. Meet in the middle, do-si-do.*

"Take the outside circle," Dad would have said. "Ride to the cattle. You've got a good traveling horse and you know the country. You have the farthest to go, so start out now and hurry." The cowboys see the imaginary maps, and they begin to move across a huge expanse of sagebrush hills and into brush or timber, riding here and there to find little bunches of grazing cattle, putting them with others. *Round you go, don't be slow, swing your partner, do-si-do. Meet in the middle, on you go.*

Without words we know who can reach those cattle first and bring them most easily, and we know that putting little bunches into bigger bunches will keep the cowboys free to look for more. We work as partners, one of us to ride to high points, locating cattle between us, and one to follow the low spots and the bottoms of the draws that can't be seen from above. Always, always look ahead and back, and look all around. *Honor your partner, honor your corner.*

The outside circle-riders bump the cattle they find toward the middle, and they watch to make sure that another cowboy rides out to collect that little gathering. *All join hands, circle the ring, stop where you are, give your honey a swing—then you all promenade with the sweet corner maid.*

My horse and I top a hill at precisely the same instant another rider does on his own high spot a mile or two away. I feel as if we're looking each other in the eye, although I can only see him skylined; I smile, imagining a courtly tip of his hat. I know who he is and what he sees by the way his horse moves. *Heel, toe, do-si-do—greet your partner, on you go.*

The rider on the hill disappears, but I wait a few moments on my windy ridgetop. Soon I see what he saw—another rider below, pushing a far-off little string of cattle walking single file, and several more crooked lines of cattle coming on different trails, plumes of dust behind them. I see how slowly, carefully the cattle are traveling and where they will meet hours later, to become a noisy, milling mass. The timing is not a coincidence—instead, it is experience, knowledge, and skill. It is artistry, a dance. *Nod to your partner, bow to your girl, greet your corner, make her whirl.*

Things don't always go right: sometimes cows will give us the slip, or someone's horse will get away, or a cowboy will get lost or bucked off, or we'll find a hole in the fence where cattle have escaped. Sometimes a bunch of far-off cattle won't join the other herd until late. At the end of the day, we'll review: How many cattle did we find? How many did we miss, and who will get them tomorrow? My friend Mike used to call it "the obituaries"—finding out what was "left for dead," she would say with a smile. "Who didn't ride far enough, or look hard enough?" When we're together, when the cattle are bunched and everyone is accounted for, the caller sings. *Then you take her home, if you find she's not flown; allemande left with the sweet corner maid, do-si-do your own; then you all promenade with the sweet corner maid, singing oh, Johnny; oh, Johnny; oh.*

When things go right—when the dance steps are precise and the cowboys honor their partners—we are part of an elegant performance. The cows and calves bawl, seeking each other, while bulls beller and paw in the dust. A glorious note of triumph sounds, the final chord rests—the cattle are together, the dancers bow. Cowboys step off the horses, loosen the cinches, and we nod in unison. "Well done. A perfect gather." We smile, offering our quiet ovation—a silent, shared respect and love for cowboys, horses, cattle, land, beauty, music, and tradition.

Honor your partner, do-si-do. Meet at the corner, on you go.
May I have this dance?

14

She'll Wear It All Her Life

We rode out in the grey chill of dawn, shivering as we watched for the first peek of sun. I pulled my collar high, tugged my hat straight, and gathered my gloves while the men set the corral gates open for the cattle to come in. Although the weatherman promised blazing-hot temperatures later today, we horseback riders waited now in a cold, blue-pink morning. As the sky brightened, cowboy silhouettes appeared across the pasture, and I heard quiet voices and nickering, snorting horses. Cows began to stir uneasily, calling their calves as though they knew this wasn't just an ordinary day.

I paused to appreciate the moment, reflecting on the past twelve months: Cattle had followed the springtime grass toward summer's mountain meadows a year ago, and then in fall and winter they trailed to the desert rangelands. Finally, they came back to haystacks and home pastures where the baby calves were born. Grass waits again in big country, but first we need good weather, good help, and good luck to get these calves branded.

I looked across the herd, and when I was reassured that the cattle were loosely bunched and that cowboys were in all the key spots, I said, not entirely kidding, "Let's get them in before we get too much help." The savvy old cows have been here before, and they hate facing this tight enclosure; they are reluctant to enter. I always imagine that they're thinking, "Nothing good ever happened to me in *there*."

If we're fortunate, we have a lot of extra help on branding day, but some of the helpers ride horses that haven't been ridden all winter; some are friends of friends, and they don't know our ways and we don't know theirs; and some are simply "over eager." If the helpers arrive with too much enthusiasm and not enough know-how, maybe bringing their dogs and their grandchildren, the action can get complicated, just keeping everyone out of the way. We'll be glad to have all these helpers, and we'll put them to use once the cattle are in the pen, but for now, easy, slow, and quiet is what we want.

It doesn't always happen that way. Last year, a rider moved to the wrong place at exactly the wrong time, conspicuously riding toward the gate where the wary old cows were already eyeing the set-up. One cow stopped, turned around, then another, and two other cowboys galloped to the gate, adding to the commotion. Somebody's dog ran yapping toward the lead. By then, all the cows were bellering and their calves were bawling, too. They threw their tails in the air and sprinted away, some of them playing and bucking, and others running as fast as they could. We spilled all of them—several hundred cattle—away from the final gate, spawning a rip-roaring wild event with horses, dogs, cowboys, and spectators racing in every direction. As a result, we had to start all over, riding to the back of the pasture and then finding a new approach to the gate.

As annoyed as I was that day, I laughed when I heard the teenage cowboys saying later, "That was great! We all got to run our horses, some of them bucked, Jim fell off, and it was just crazy! We had a blast!" What *I* saw, though, was the time we'd lose, and the chance someone would get hurt, and the possibility a calf might run clear back across the creek and through the fences. I laughed later, too, when an old cowhand told someone, "Just like a goddam salami, when all those cattle came boiling back over us! Not a goddam thing we could do but watch 'em go." (Later, it dawned on me. He meant *tsunami!*)

This year we closed the gates behind the cattle uneventfully. Satisfied and relieved, cowboys nodded to each other. "Well done. Just right." Every minute counts.

Now, we're glad to see the extra help come wheeling in—pickups and big trailers full of horses, jolly cowboys, neighbors, and friends. We haven't seen these folks all winter because they've been busy and so have we. The mood turns festive.

"Have you branded?" cattlemen ask each other. "When can you turn out? The grass is ready!" Ranchers don't turn cattle out to open range until the baby calves are branded. In Wyoming and other western states where ranching still hangs on, a permanent brand is as good as a passport, giving identification to that animal forever. Once it's branded, a cow or calf can always be traced back to its owner, even if it has strayed afar.

"Tell me what to do," our daughter Carol calls out to Tim. "Give me a job. I can vaccinate or brand. *Please* don't make me wrestle calves anymore!" She and her family don't live at the ranch, but they come to help each year, to share the day. Tim laughs and hands her a branding iron. "Okay," he teases, above the noise of the cows. "You're a senior citizen now. High school kids *like* to bust calves. We'll let them do it. They're better at it, anyway."

Tim hands out the other jobs. "Mart, Scott, and Cindy, you rope first. Trade with others, off and on. Mom, Dad, and Carol will run the irons." He puts the wrestlers into pairs and he names experienced helpers for vaccinating, doctoring, and other jobs. "Be careful," he cautions. "Watch out for each other. Pay attention and nobody gets hurt." No dogs are allowed near the cattle, no sandals, and no little kids unsupervised.

"Irons are hot! Ropers, bring 'em!" Stan declares. The ropers ride into the herd, purposefully swinging their loops. Within seconds, they drag calves toward the calf wrestlers, who reach out to lay the bawling calves flat on the ground; the ropers pull past, horses leaning into the weight. Different regions throughout the

West have their own techniques and traditions; at our place, the cowboys catch the calves by the two back heels and drag them in carefully that way. Other places, the ropers catch the calves around the neck. Calf busters and ground workers don't approach the horses or start grabbing the ropes or calves until the roper signals that it's safe, and when the ropers are catching "fast," they too remain outside the action, waiting a turn to bring the calf through without stepping on the busters, some of whom may still be on the ground.

"Coming your way," the roper calls. A swarm of workers carrying red-hot branding irons, knives, and vaccine guns step nimbly between calves and ropes, alert to every scenario. They keep an eye peeled for a protective mama cow or a skittish horse, for sick calves, for any unusual circumstance. It's all part of the job.

Teenage calf busters (girls *and* boys) offer wisecracks and laughs, showing off their muscles and know-how. They coach smaller kids and teach them tricks for safety and efficiency. Our daughter Sara's freckle-faced boys, Silas and Pete, are here this year, eager to learn—gaining confidence by the minute.

"Come with me, Pete," cousin Marshall says. "I'll show you. Sit flat, grab the hind legs, and take the rope off. Lean back. I have the front legs and the shoulder. Front guy turns loose first. Understand? Front guy. Otherwise, somebody gets kicked. Do it right." Calf busting is dirty work for sure, with dust, mud, and manure everywhere. Brute strength helps, but timing and balance make a big difference in keeping the calf quiet so the brand can be applied straight. "Over here, Silas," a husky teenager calls. "I need a partner."

Old traditions and protocols developed through time keep the calves safe—as well as the helpers. The jerk of a rope on a calf's leg could cripple him, make him worthless at market, not to mention causing him pain. If a roper is too rough, the boss might ask him to slow down or "take a break." Someone walking carelessly through the melee can trip over a taut rope or be snared or tangled. The horses won't step on anyone by choice; they're savvy about the confusion, more so than many of the helpers. It's everybody's responsibility to pay attention.

I automatically count the helpers. I fixed a mountain of food yesterday, and I hope it will be enough. There's plenty, I decide, knowing that each family brought a dish to share. "Those teenagers can put away the food, though," I remind myself. I haven't forgotten the carload of college friends Dan brought home with him to help one year. The weather turned bad and rain poured the entire weekend. The "boys" played poker in the bunkhouse, ate all the food I'd prepared ahead of time, drank all the beer, and drove cheerfully back to the university in Laramie. "Thanks," they called as they pulled out of the driveway. "We had a great time." We had not branded a single calf. Adding insult to injury, one of the boys offered me a recipe of his mother's, which he thought I might like to use next time. On Monday, I started all over—meat, potatoes, salads, rolls, cakes. And, of course, I had to buy more beer.

Carol, Stan, and I work steadily, placing the brands precisely on each calf. "Put it on right," I was taught. "She'll wear it all her life." Our brand is the left-rib Diamond Tail, registered to Stan's grandfather in 1906 and used here without interruption since that time. In the State of Wyoming, shoulder, hip, or rib brand locations are approved and registered, and no duplicates are allowed within the state. The Diamond Tail is a diamond with an attached bar—the tail—pointing out to the side.

Putting a brand on straight is tricky, and it's a point of pride for me that I'm good at it. Stan's dad taught me—I look carefully to place the tip of the diamond first, behind the shoulder, and I roll the iron back until all of the hot metal is firm against the calf's hide. This is a rib brand and needs to be high enough to be read from "a-horseback" and low enough to be seen by a man afoot. It must be straight, with the bar horizontal. I hold it in place, and I don't let it slide. I don't want to take the teasing, either, if I place a "dragging" tail or a "flying" diamond when the brand tilts up or down. (Simple "one-iron" brands like this one are sought after because they're easy to apply and easy to read; a complicated, designer brand like "Triple Heart Triangle" is all right on a gate-post or a coffee table, but useless to stockmen.) "Running the iron" requires an attentive, steady eye, and the job comes with prestige attached. No one grabs a branding iron without being asked to do so. We don't want a blotched, unreadable mark on the calf. At our place, someone in the family must run the branding iron.

The smoke rolls black and smelly as the iron singes the hair, and a teary little girl worries about the calf. "Don't worry," I tell her. "It only hurts for a minute, like when you touch something hot. She'll be fine."

In springtime Wyoming, lots of newspapers do a branding-day feature. Photographers gravitate to the colorful spectacle, as do artists gathering material for a next painting. We don't mind. We're proud to share the day, a festival of sorts. I'm still surprised at how many of our rural, small-town neighbors know so little about branding, or cattle, or agriculture in general. Today, our local newspaper editor (who brought a cake) snaps pictures every-where: cattle, horses, rugged-looking cowboys in Stetsons, smoke and dust rising above pole fences, gorgeous girls in tight jeans, silver spurs, coiled lariats, little kids in big hats and Wranglers.

"Stand still just a minute, will you, for a picture?" he asks of Carol. She is slender and fit, lovely in a brimmed hat, jeans, and well-worn boots. Behind her, cowboys' weathered faces and beat-up hats make classic portraits, too.

"It seems like yesterday Scott was a little calf buster," I say. "And look—now he's got *grandchildren!*" Mustached and handsome, Scott is an artist with his loop; he seldom wastes a throw, and he makes no commotion in the herd. Most ranches pronounce that nobody ropes without a nod from the boss; roping's a high-prestige job, as it requires a sixth sense about safety. "Practice roping at home, not here," is the rule. "And no broncs. Get your horse broke before you bring him." Ropers take turns, three or four at a time at our place, rotating in and out to keep a steady pace for the rest of the crew. Scott's young grandson is already skilled and careful, thrilled when he's asked to rope a few calves.

Hours pass. Jackets that felt good this morning are draped over the fence now, and some of the high-schoolers are working on a sunburn, determinedly wearing T-shirts and ball caps instead of hats and long sleeves. Workers methodically, rhythmically handle the calves one by one until every calf carries the Diamond Tail brand. By late afternoon, we and our helpers have gathered, sorted, roped, and branded several hundred calves, and finally we count the cattle out the gate.

The hot, sweaty workers head to the lawn for frosty-cold beer and lemonade; cowboys take the horses to tie them in the shade, loosening the cinches. Some of my friends—what would I do without them?—have watched the food table and the kitchen, stirring, heating, and carrying as need be. There's a bucket of soapy water so the cowboys can get the grime off their hands, and then they grab a beer or lemonade; soon they heap their plates, choosing from enormous roasters of beef and pinto beans, tubs of salads, trays of home-made rolls, fruit, and desserts.

I pop the top off a cold drink for myself. My feet hurt, and my back is stiff from lugging the iron. I'm dirty, the sleeve of my shirt is ripped, and my Levi's are splashed with manure—definitely not photogenic. I'm ready to call it a day, to relax as we congratulate ourselves on getting the job done. I haven't had time to talk to my friends yet, and I'm especially eager to greet the cowboy elders who leaned on the fence most of the day. These veterans of many, many brandings proved long ago that they're top hands. They curse their gimpy knees and arthritic hands; they grin and poke fun at each other and the "youngsters" who are pleased to get a nod from this exclusive circle. "Come on, Bob, there's a chair over here," someone calls. "Damn," old-timer Bob says, "if I sit down I might not be able to get up."

"We're glad to see you, Bob. I learned all I know from you. Well, half maybe," someone says. Others add their own affectionate teasing. "Looks like a man of your stature ought to be able to afford a new hat," one of them says. "Savin' my good one," Bob replies. I silently acknowledge other friends who used to be here, those who have passed on, but I don't say anything out loud. I get a lump in my throat thinking of Carol's husband, David, whose last outing was here with us on branding day, only a few weeks before he died of pancreatic cancer. A lighter memory is of Steve, our longtime partner; I remember how he always said "I'm not using that flimsy fold-up chair—it's like sitting in a goddam bucket."

A bunch of high school boys walk by, and I move to thank them, saying, "I don't know all of you by name, but I sure want to say we appreciate your help. Without good calf busters, we couldn't do this." One boy, startled, takes off his sunglasses and says, "Mary, you know *me*. I'm Riley!" He lives a few miles up the road, but I haven't seen him for a while. He towers over me now, as do his friends. We all laugh and hug, figuring out who's who. I realize that I do know most of them, that they've just "gotten past me."

Carol and I fill our plates and find a place to sit together, attacking favorite foods. "Mmm. I'm starved. Great potato salad. Did you try the rolls? Who brought the coleslaw?" She is in her early fifties; I am seventy-some, and today we happily, lovingly reminisce about previous brandings, talking of grandparents, families, and others who understand this tradition. "Last year we were short of help; today was just right—plenty hot, though. Remember the year we got *snowed* out? And when the high school kids all got the major sunburn, with the *Prom* happening that night?" Everybody has a story to tell: when a rookie barged in to rope and then let his rope get under the tail of another guy's horse; or when somebody got jabbed accidentally with a vaccine needle; or when a girl's skin-tight Levi's split out at the seat. "Em-Bare-Assed, she was," Bob says. "*And* embarrassed."

Cowboys stretch out in the shade, talking of cattle prices or hay supplies, making jokes and sharing family news. Teenagers start a ball game and little kids begin a water fight, shrieking and throwing ice as they run across the lawn. We laugh and move aside. "We got it done again. Let's try the pie," Carol says, and we climb stiffly to our feet, heading down the trail to next year.

15

Luck and Miracles

During calving, Stan and I are often horseback together, riding each morning through the cow herd and the calving pasture. The routine includes troubleshooting if there's a problem with a cow or calf, but it's usually uneventful—a pleasure ride. One Monday, we had gone across Shell Creek to the east end of the ranch toward a far-off meadow, a lovely place for calving because its trees and brush offer cover for the new babies when the weather is inclement. We frequently ride the gentler colts, to give them some experience. Stan rode the young bay Whiskey Blizzard, and I rode a sorrel gelding called Shooter, both only green-broke but quiet and smart.

We were busy that morning, but nothing unusual. It was chilly with a skiff of new snow, patches of ice in the shade, lots of new baby calves. We eyed the diminishing haystack and talked casually about the price of hay and the cattle market. As the bright sun warmed things up, we remarked about the goopy, slick mud starting to suck at the horses' feet, and the ice beneath.

For the last circle we parted company; Stan wandered one way and I another, enjoying the sun's warmth. I rode slowly, whistling a tune and daydreaming while I watched the little families of cows and calves. Birds were chirping, too, congratulating each other on their return from the south.

As I turned toward home, I thought I heard something odd—a yell, maybe. I listened for a moment before deciding I had imagined the sound. I kept moseying along and then, "Help! Hurry!" Again, for sure: "HELP!" Stan's voice.

I hurried back toward the sound, toward where we'd parted, but I couldn't see anything amiss. No horse, no rider. I couldn't see the dog, and I didn't see a disturbance in the cattle. There! Now I saw the dog standing by a ditch, almost invisible at the corner of the field. I galloped toward him, and immediately I knew we had big trouble. Whiskey was down on his side on the edge of a ditch and Stan was pinned under him. Neither could move.

The greasy, black clay had softened into slimy, treacherous mud. When Stan rode across the little ditch, Whiskey didn't take it seriously enough to jump. He stepped right across it at a walk and slipped down flat on his side. The horse was laying with his feet uphill, his back downhill, and Stan was underneath on the downhill side.

"Good God. Are you all right?" I asked.

"Yeah, hurry. I'm pinned. He can't get up."

Hurry? Do what? I climbed off my horse, afraid to make a move—what if Whiskey spooked, what if he struggled? Stan was under horse and saddle, and the saddle horn was close to his chest. Completely aware that the horse was young and unpredictable, I eased over, talking to Whiskey, "Whoa, buddy, whoa, take it easy, quiet there, take it easy." What to do, what to do, what to do . . .

Stan's leg was already numb. I carefully unbuckled the cinches and breast collar and freed the saddle from every strap; at least if the horse ever could stand, Stan's foot would not be caught in the stirrup. What next? How could the horse get a footing to stand up? What to do, what to do . . . if Whiskey moved at all he'd roll right onto Stan. No other way he could move, no place he could go.

Of course I couldn't lift the horse or pull him. His head was downhill, too—should I go for help? No, bad idea. If I rode away, Whiskey might panic or thrash around, struggle—God knows what. My horse's presence might keep him calm. Whiskey had not moved a muscle; he and Stan both lay completely still. Stan's eyes were closed in pain; he offered no suggestions—except, "Hurry." My horse stood near, nibbling at the grass nearby.

I took my lariat off my saddle, trying to figure out any solution. Could I pull Whiskey free with my own horse? Would that make things worse? I wondered again if I should go for help. No, better not leave. To see what would happen, I carefully eased the rope around Whiskey's back foot and tugged lightly, wondering if I could pull the foot forward under the horse to give him footing. It must have occurred to Whiskey that he could move then, because he gave one huge lunge and jumped clear up. He didn't scramble—just suddenly leaped up—all the way up!

Somehow he didn't step on Stan. Somehow, somehow, he didn't roll backward or downhill. Somehow his hind legs grazed by Stan's shoulder and the saddle, and there he stood, leaving Stan in a heap on the ground and the saddle and blanket, too, untouched.

Stan sat up and I sat down. I put my head in my hands, trembling and breathless. "I'm all right," Stan said. "Nothing broken." Whiskey shook himself off, shook again, and lowered his head, his one side a plaster mold of black clay. When I stood, I rubbed the horse's ears, scratched his mud-covered face. "You sweet thing" I said to Whiskey. "Kind, gentle thing. Good Lord above."

Stan swore at the pain as the circulation came back into his leg, and after he caught his breath, he said, "My own fault. It was a bad spot. I knew better; I wasn't paying attention." We resaddled Whiskey, mounted up, and headed for home. A slow quiet mile or two, thinking "what if."

When we got to the barn, the mud was dry and caked on Whiskey, Stan, and the saddle.

"Have some trouble?" somebody asked.

"Yeah," Stan answered. "We were lucky."

I hear myself saying "good luck" when cowboys or friends or families part. To me it means "be careful, wherever you go" and "pay attention" and "hope for miracles everywhere." I think about luck each time I see a near miss, a close call, or an accident that *didn't* happen because someone was paying attention. Someone with experience or calm or quick reflexes stepped in at the right time. A good, quiet horse didn't explode when he could have; a gentle colt like Whiskey trusted and waited and did not panic.

Days are generally filled with more good luck than bad, and more miracles than we stop to count. At each day's close I give thanks for things that turn out right, no matter what we call them.

My journals often tell of weather, pasture changes, or family occurrences, and they sometimes make vague acknowledgements of ranch happenings. The brief entry for this day, though, brings a shiver, and as I reread it, I relive that close, close call.

> March 20, 2008:
>
> I didn't talk about this for a while or write it down. It was just too scary to think about. It was either incredibly good luck or a miracle. Maybe both. Maybe they are the same thing.

16

My Ranch, Too

One September afternoon I drove a heavy livestock truck down the steep, winding Shell Canyon highway—carefully, but not anxiously. Snow was predicted, which would bring ice to the road and mud to the corral where we'd begun our day's work, but here, the highway was dry. The truck was loaded with noisy, bawling, big calves, just weaned—a full load, and I paid attention; shift down, save the brakes, take the turns wide and, especially, "don't get in a hurry." In my concentration, I'd almost forgotten that a woman sat beside me, until she spoke. "I envy you," she said. "Things are so much easier for you—you have so much *confidence*. You know how to *do* things."

Her husband, a neighbor, had offered to haul a load of calves for us, too, so we could get them all down before the storm set in. His truck was behind us, and Stan drove a third one in our caravan. The woman had volunteered to ride with me, although we didn't know each other very well.

"That's not the way it works," I said, preoccupied, and a little annoyed, thinking to myself that things aren't the way they look. "I didn't always know how to do things like this," I told her. "It doesn't have much to do with confidence." After a moment's silence, I wondered if I'd been a little sharp, and I went on to tell her about the first time I drove a truck on this canyon road. "I was scared to death. The men asked me if I would do it, and they told me it would be just that one time, if I could '*please,*' they said, it would help so much, and they would keep a close eye on me, they said. I fell for it, and I got behind the steering wheel. I didn't have a test drive or lessons at all—just 'don't ride the brakes'—and away I went. That 'one time' turned out to be a joke. I've been driving the canyon ever since, I guess. Through the years, someone often asks with a grin, 'Can you drive truck when we're hauling cattle? Just this once?' We all know I'll be asked, again and again. But I just go slow, take my time, and I don't get in a hurry, and so far I've made it every time."

"We'll be past the worst of this, in a minute," I told her, barely touching the brakes, shifting down, and finding a lower gear as we approached a "steep grade" sign—one of those bright yellow ones showing the black shape of a truck, comically nose-diving down, down, down. "I don't mind driving. I like learning to do new things. And I like helping and seeing that the work gets done. It's my ranch, too."

Young women ask me now, "How did you do it all? The physical work, and being out in the weather? Practically being a hired hand, and taking second place to men? Raising a family, being a mom and wife? Would you do it over again? Did you ever wish for a career of your own? Did you ever want to do something else?"

Those questions make me smile, because when I think of interesting, admirable women in the West, I remember classic memoirs such as *Letters of a Woman Homesteader,* by Elinore Pruitt Stewart; *Cowgirls: Women of the American West,* by Teresa Jordan; *Lady's Choice: Ethel Waxham's Journals and Letters, 1905 to 1910;* and *Hell on Horses and Women,* by Alice Marriott. These and other accounts of pioneer women tell of nobility, courage, and determination—and strength and resilience. I didn't see myself in that pattern, only just doing what I could do, probably much the same as they did, and as my mother, my grandmothers, and other women I remembered did, too.

To answer these women's questions, I tell them, "Things change. Times change. It was as different in my own day compared to yours, as it was between myself and the older women I knew." To explain, I first try to describe the women I most admired.

When I was a little girl, my grandmother Lula McGinnis Budd was my heroine—a strong-minded, smiling woman—worth imitating. She was only a toddler when her father, William McGinnis, brought his wife and children to Wyoming Territory. He had joined the Union Army in 1862 at age fourteen to fight in the Civil War, and he had mustered out in 1865 without ever having seen battle. He went to Missouri and then to Utah, where he found work in the silver mines in Park City. He settled with his family in Wyoming in 1886, moving to an abandoned homestead on the Lower Green River.

At that time, the family, which included Lula, "brought with them fifty cows, four horses, a wagon, and $500, camping along the way until they found some land that would make them a living," as she later wrote, in spidery handwritten notes, for a local historical society. "A few things I remember, however it's possible they are happenings that were talked of so much so many times it impressed me as I feel I remember them," she continued. "Father had worked in the silver mines in Utah for almost 25 years, and they both wanted to get a piece of land and have a ranch and cattle."

I romanticized her childhood, though, and I once said to her, "Oh, I wish I'd lived in the olden days." She chuckled softly and shook her head, cuddling me against her side. "No. No, you don't wish that. It was awful hard." She described sharing the family's small supply of food with the pitiful, hungry Indians who passed by. She talked of the woeful attempt at a vegetable garden, of trying to stay warm during the winters—constantly chopping and hauling firewood. Grandma told me how families sewed and mended their clothes, washed on a washboard, and hung their laundry on a line in freezing temperatures. She made no mention of men helping with the housework. I assume that varied for each family; presumably the men were working hard, too, with livestock out-of-doors.

Grandma's historical notes described other hardships: "The most tragic thing that ever happened to us was when our little 18-months-old brother drowned in the Spring Creek. I remember so well the men making the little casket, it was just rough lumber. Then the women lined it with white and covered the outside with black material. Our dear friends worked so hard to have it nice."

"When the Indians came they were always friendly," she wrote. "They had us marked for a place to eat I guess. They always asked for hot biscuits. Mother always made things for them, whatever she had on hand. They always took it outdoors to eat. There was one I really liked he always brought me beads and held me on his lap. One time they came and wanted a sack of flour they traded 2 horses to Father for 1 sack of flour. We named them Punch and Judy. Judy never did get tame but Punch was the joy of our lives; he never cared how many of us got on him and we always had him handy when we were playing."

Their first ranch proved to be unsuccessful, so at a spot nearby, later known as Midway, they built a "road ranch," selling meals and lodging to travelers and pasture to ranchers who drove live-stock to the railroad. Lula wrote, "Mother got the post office there at Midway, and she was proud to say she served under three presidents. The mail stages stayed overnight at our place, with all their passengers usually 1 to 4." She told me, "I hated that. Nothing but work. Cleaning the oil lamps was the worst."

Grandma met John Budd at Midway—"a handsome young cowboy, he started riding through there," she wrote. She was twenty-three years old, Grandpa was twenty-seven, and they married in 1905, after Grandpa built a sturdy log house and headquarters for them at his homestead on North Piney Creek. Grandma Lula had watched the progress of her own family and the people around her, and she knew that she and Grandpa would have to work hard to "get ahead." From the beginning she shared ownership in that opportunity: it was her ranch, too, and she worked as hard as Grandpa, I believe, to make a success of it.

My mother, Ruth Peterson Budd, was a "town girl," but no less a ranch partner, as time would reveal. She grew up in towns along the railroad line as she, her sister, her mother, Winifred, and her father, Charlie Peterson, followed his job as a storekeeper for the Union Pacific Railroad. Mom attended Loretto Heights College in Denver, where she studied violin and learned to speak French and Latin. She took a school-teaching job at a country school on the Green River near Big Piney, Wyoming, and she chose a young rancher, Joe Budd, for her husband-to-be, although there was no shortage of eligible young men. When Dad announced their plans to marry, Mom's family was shocked. "Joe," her mother said, "you're a fine young man. But Ruth is a *town girl,* and she's tiny, besides! A *lady!* She can't live clear out there on a ranch." According to the family story, Dad replied, "Oh, Mrs. Peterson—don't worry about that. I'll take care of her. She won't have to do a *thing.*"

They married in 1937. My dainty little mom learned to ride horses, handle cattle, pack salt, fix fence, carry firewood, drive trucks, and cook for ranch crews. She could do just about anything a situation required. In my favorite childhood memories, I see her standing in tall mountain grass, sipping coffee from a thermos lid while Dad unlashes salt blocks from a pack horse; I see her always attentive, anticipating how best she could help. Sometimes when we sorted cattle, she sat horseback at the edge of the herd, keeping meticulous tallies with a tiny pencil and notebook. At home, she kept a little repair kit of screwdrivers, pliers, and wrenches, saying that she didn't have to "wait for a man to fix things!" She wore baggy pants and a droopy cardigan around the house—a Camel cigarette and a Zippo lighter handy—but when she dressed up, she was astonishingly beautiful. My sisters and I loved to see Mom and Dad at a dance, where she twirled in a pretty skirt, her lovely, slender legs set off by nylon stockings and high, high heels.

I don't know much about how Dad's parents made room for the "town girl," or when Mom came to feel that the ranch also belonged to her. Dad told me that in the early years of their marriage, they stayed summers in a mountain cabin but lived with Grandma and Grandpa Budd during the winter. "I wouldn't wish that on anyone, sharing a house," Dad said. "But those were hard times, and we didn't have any choice if we were going to get ahead."

Ranch women of that day did not participate in town activities much; as Mom and Grandma both would say, "A ranch doesn't stop for me. I just can't get involved with things like that. People would be counting on me for cards or something, and a crisis would come up at the ranch, and I'd have to chase off after tractor parts or run an errand of some kind." (It seemed to me they felt these jobs to be important, not *un*important.) Still, when she had reason to be in town, Mom found time to stop for coffee with Dad's sister, Helen. I don't know what they talked about, but sometimes the ashtrays overflowed while they giggled and chattered. Other times, we kids were not supposed to notice if their voices were very low or if one of them frowned. I believe their own marriages and those of their friends probably held hurts and trials, just as man-and-woman relationships and families of today do, and I imagine that as friends they confided in each other and leaned on each other. Those were the 1940s and 1950s and, for the most part, Mom's credo seemed to work: "Make a clean place to work, however small. Get some fresh air. Get some exercise. Keep busy."

Women like my mother and Aunt Helen didn't sprout from family trees, but instead they appeared by rising to circumstance and challenge. Stan's mother, Maureen, was born in Ireland, in 1903, and arrived in Butte, Montana, with her parents, Timothy and Anna Desmond. (Rare for that day and age, Timothy and

Anna eventually divorced, and then Anna raised their two girls alone, making a living as a nurse during the influenza outbreak in Montana.) Maureen and her sister both attended college in Missoula, Montana, and Maureen took a teaching job in Greybull, Wyoming, where she met and married rancher Howard Flitner, my husband's father. Maureen was a lively, fun-loving, independent woman, vivacious and pretty, and was said to have scandalized the school in Greybull by pronouncing that girls in physical education classes could wear "bloomers" instead of skirts! Maureen, like my mother, "married into it" as the saying goes, when she entered the ranch world as a wife. She didn't work outside the home or take a big part in physical ranch work, but I knew her to be a dedicated partner, truly loyal to her husband and his calling.

Stan and I met in college, and when we married, my new place was three hundred miles away from my childhood home, so I only had to move from one ranch to another. Agricultural economy had already hit a downturn; it seemed natural for me to pitch in however I could, and I liked it that way. I don't remember conversations about gender equality; I grew up thinking that ability was one's defining factor, and in our upbringing, if my sisters and I could learn how to do something, we did. My married life let me continue in that direction. I liked the challenges of outdoor, productive work, and small, visible achievements. I knew how to ride, and I took every opportunity I could to do that. I did a lot of the bookkeeping and message-taking, too, and I fed extra people who showed up at our table. I hurried to town to get repair parts, I cleaned the tack room in the barn, I helped feed livestock, and I filled in wherever I could. "Just trying to get ahead," as my grandparents had said. Our children followed those patterns, learning from us. "We've all got to pitch in, if we're going to get ahead."

My young women friends have trouble following that train of thought now; it's hard for them to view a wife's role as helpmate, back-up, housewife, and mother.

"Marriage? It began with love and romance, one hoped, and it offered security, a family of one's own, a community, a place. A way to get ahead," I tell them. "Things were different, remember." It seemed that most women didn't express their dissatisfaction beyond a little harmless complaining at first. Gradually, we explored vocabularies about self-worth and fulfillment, as identified for us by author Betty Friedan in her 1963 book, *The Feminine Mystique*. I did not see the legions of unhappy women that Friedan described, but perhaps ranch life had left me apart from them. A generation or two earlier, I imagine my grandmothers were too tired to ponder those topics, and whatever unhappiness or loneliness they did suffer, they didn't share. My mother, if speaking to a feminist friend, would likely have said, "Get some exercise. Get some fresh air."

"At the time," I explain, "I didn't spend much time thinking about it. My sisters' and my friends' lives paralleled mine, for the most part. Ranch life offered variety and challenge and kept me busy—and I felt I was a team-mate in a demanding, interesting profession."

Our house was full of kids: our own, our nieces and nephews, and their pals. I wasn't bored. Stan and I had lots of friends—interesting, stimulating, intellectual women and men, who liked being at the ranch and sharing days with rowdy cowboys and travelers. Sometimes, we rallied for dinner or picnics or drinks, and a chance that—spur of the moment—we might dash off to the local bar and grill that offered live music. We were on the go constantly, juggling work and play.

After the ranch property was divided between Stan and his brother, my voice strengthened as Stan and I actually began to ranch for ourselves. At the same time, my role was still back-up and helpmate, and that was fine with me. I knew that my strengths in accounting, organizing, and problem-solving were as important as the heavy lifting Stan did, and he recognized that, too. I

didn't have the requisite knowledge to make management decisions about pastures or cattle or crops. But I was an equal partner.
In our marriage, Stan and I had plenty of disagreements, but they
never resulted in any kind of violence beyond one of us angrily
slamming a door or stomping out of the house. Around us we saw
alcoholism, infidelities, divorces, bankruptcies, and tragedies that
affected whole neighborhoods, but drugs were not readily available, and X-rated media had not reached our community.

In the large personal and business decisions, Stan and I usually found common ground. It helped that our parents had lived
according to similar moral and ethical values and that their politi-
cal, religious, and social structures were very much alike. Both
families were Catholic; both were hard-working ranch people;
both were interested in local schools and communities.

I wasn't a martyr. I lived the life I chose, at home and on the
ranch. I found friends who were avid readers, as I was; I joined
book discussion groups and took night classes at the school.
I belonged to humanities groups, and served on the Library
Board, a museum board, the School Board, and local and state
government committees, including the Wyoming Game and Fish
Commission. I loved working with the local 4-H club. Later on, I
participated in an agricultural leadership group that took me to
Canada, Washington, D.C., and South America. I had to juggle,
sometimes, to make things work. I remember that when I returned
from a two-week trip to South America, calving had started, and
on the muddy afternoon I arrived at home, I immediately put on
my coveralls and overshoes and went straight to the barn! I didn't
unpack my suitcase for a week; I didn't have time and, for sure, I
didn't need any of my dress-up clothes.

My journal recorded the sadness of accepting the deaths of
my parents and Stan's, and saying that same goodbye to many,
many good friends. In our hard or sad times, I wasn't stoic, either,
although I wanted to be. I wrote of watching helplessly as our family

endured accidents, illnesses, and sorrows, as ordinary people do. Fears and worries are recorded in my journals, jotted on the pages beside notes on the weather and the business of each day.

Then and now, I find that balancing the roles of mom, wife, and business partner is the most difficult challenge. The relationships are never "just business." We used to laugh at a slogan that described our power structure: "A dollar doesn't wait on a dime"— reminding us to be ready for what might come next, in any day's work. "Time's a-wasting!" Of course there can be confusion about who's got the dollar and who's got the dime. Petty disagreements bring angry words, words that can't be unsaid. Often, later, someone apologizes or says, "I didn't mean it," but I fear a time when someone will go too far, say too much. I don't want to "take sides." Fortunately, love, loyalty, and trust among family members have so far remained sacred in our big pictures of money and ranch management, held together by our common allegiance to the land itself.

I find that women and mothers try to guard against the tumultuous relationships between fathers, sons, and brothers, and we instinctively sense the dangerous human triangles described in age-old dramas or Greek mythology. Hera, goddess of women and marriage, guided, protected, or manipulated the behavior of her wayward husband Zeus and his offspring, for better or for worse.

"It wouldn't be honest," I say, "to pretend that my life on a ranch has been idyllic, or that I was happy every minute. But a bad day is completely different from a bad life, and I have a good life."

"I'm not sorry I put my home life first," I tell my young friends, "because the ranch was part of that. Without the physical and emotional challenges of ranch life, I would probably have been more restless, especially after the children were in school."

Certainly, I sometimes felt bullied or dominated by the men around me. Sometimes I felt unappreciated, angry, or hurt, or left out—but most of the time, I didn't feel that I was vying for control or authority. I think I felt then as I do now: that people are people. Some of them are men and some are women. Some give respect; some earn it. In my own life, I was fortunate to have opportunities—and to live in a setting—that allowed me to earn respect.

I observed, however, that most men don't like taking orders from women, and lots of people don't like taking orders from anyone. As a leader or as a boss, getting past that form of disrespect is never easy, for man *or* woman. A cowboy helping here at the ranch ignored me when I asked him to step aside, and I had to tell him again: "Please. I'll do that myself." He answered, "I know what I'm doing." "Maybe you do," I said, "but they're not your cows." He yielded reluctantly, but he pouted the rest of the day. Like some men do, maybe he thought his gender gave him superiority, but I didn't see myself as a woman that day; I was the owner of the cattle, the boss. I thought he should see me the same way. I tell my young friends, "I just try to ignore it." It is tiring—to hint and strategize to get people to do the work *my* way. The language "as good as any man" doesn't say much to me, although it is a vocabulary that some present-day cowgirls use.

During the up-and-down ranch economies of the 1960s, 1970s, and early 1980s, women took on new roles: ranchers struggled again for solvency, and many ranch women shouldered part of that load in a different way than Grandma's generation might have. My women friends boosted ranch income by finding jobs in town—as waitresses, sales clerks, and receptionists, or they worked at the local schools. Surprised at themselves, they said things like: "I love it! It's easier than the good wife, the good hired-hand job! It's *heaven.* I go to work at the same time every day and come home at the same time every evening *clean,* and I don't have to jump every time my husband changes the plans on the ranch."

Stan and I did that math for ourselves. In our case, if I got a job in town, we would have had to hire a man. I could work better, smarter, and cheaper than most men we could hire, even if

Stan had to do more than his half of the heavy work. I liked being close to home. I didn't really want to get a job in town. Anyway, the only jobs that actually paid women as professionals required education—schoolteachers or the occasional attorney, perhaps. After Stan and I married, I never finished college—common at that time—so I had no real credentials for a high-paying job.

A couple of times a month, we paid a woman to help with the housework, and we got by pretty well that way, most of the time. Definitely, I did my share of the worrying. I felt that my self-respect and the respect I wanted for our family depended on our success as ranchers. I wanted to "get ahead." Today's ambitious, capable women find opportunities for executive positions and high-profile jobs, and they seek career paths instead of only a paycheck. These smart, motivated women now find it acceptable and possible to take on ownership or management roles in agriculture, too, studying for advanced degrees in areas that were once limited to men. My own two daughters both have business careers, as do my two daughters-in-law. "I think you can do the things you want to do," I tell them, "just maybe not all at the same time. You can identify your priorities, and make your choices accordingly. It doesn't have to be either/or now."

Recently, a woman who had committed wholly to ranching described to me the frustrations of her ranch marriage, and we talked of disappointments that occur within such a complicated relationship. She is well educated, and with great difficulty she juggles a different career beyond the ranch and family. In that "behind-the-scenes" role, she works endlessly toward the needs of their ranch business, with great emotional and physical toll. "I don't know how long I can keep doing this," she said. "Or why I stay," she added angrily. She told me about the latest in a series of insults and about the visible lack of respect her menfolk have for the value of her time and participation. "I'm just so tired of it. My husband and I love each other—I guess we still do. How would I know? There's no time to talk, except about business. The work never ends. Everyone's always worn out." Suddenly her embarrassment at sharing this confidence stopped her short. Her face flushed and she waved her hand, pretending to brush away her problems.

"What will you do?" I asked. "Do you want to leave?"

She smiled a little and shook her head. "No. I won't leave. Sometimes I think about it. But I won't. Because it's my ranch, too, you know. Family, livestock, land, dreams. I didn't really plan it that way, but here I am, my whole life invested in it. I was a damned martyr and I put up with aches and pains and hurt feelings, and my husband bullied me and the boys, and I'm mad about that now, and I wish I hadn't let that happen. But at least we still *have* a ranch and a family, and if I left, we'd have nothing to share without the land. I'll never leave. And anyway, I love this ranch. It's part of who I am, even if I hate that phrase!"

I understood. Without saying more, I compared her life to mine, and I thought of Mom and Grandma and others. I didn't carry water in a bucket or use a washboard or cook on a woodstove as they did, but often when I came into the house at night, I was tired to the bone from handling baby calves or lambs, lifting bales or riding horseback. I worried constantly that the children worked too hard and that they took on too much responsibility for their young souls to bear. I thought of the children's school activities I missed, arriving late and dirty sometimes, weakly apologizing, "We didn't finish at the ranch in time." I fretted over the bank accounts. I was afraid for our family's security, afraid we might fail in business, and afraid I was not strong enough to keep on going. I believe the strong women who went before me had those concerns in common, too. They wanted to "get ahead," as several of them said. "Make things work." They loved the land and the life, it seems, in many different ways.

In my journals, I didn't write much about ugly, personal moments of our marriage, our partnership, and our ranch life.

Early on, a grade-school teacher taught me, "Never write down anything you don't want to share with the world." I took that to be good advice after she read aloud to the entire classroom my written note to the boy across the aisle, telling him that I would love him forever. I can read between the lines, though, and if I wanted to, I could search out old wounds, leafing through the pages, remembering details of hurts or sorrows that marriage naturally holds. I see no advantage to that, as each life is different; each joy or sadness is unique. I prefer to move forward.

Being a ranch partner is not everyone's choice, nor should it be. Ranching is my profession, a place where I have contributed and shared in achievement. Occasionally, I think of the woman who rode with me in the truck that day—recalling the words she said. I don't know what she sought, or what she saw, but I think that "confidence" comes with finding a place of self-assurance and satisfaction.

It worked out that way for me and for some of the women I admire the most, here at "my ranch, too."

17

Home Sweet Home

During the ten years we owned a ranch near Moneta, we trucked the cattle to and from Shell each spring and fall, 150 miles each way. One fall when feed was scarce at home we hauled the cows to Emblem Bench, near Greybull, where we bought some beet-top pasture from Werbelow Brothers' farm. That feed would last a month or two.

When the pasture ran out, we considered how best to get the big pregnant cows home. To trail them meant 25 miles for the cattle to walk—15 miles with horses and riders on Highway 14 toward Greybull and then another 10 miles on back roads to the ranch. The alternative was to truck them—expensive, and also stressful to the cows themselves, heavy as they were with calf. We opted for the trail drive, even though serious winter weather threatened.

We started toward home one New Year's Day in freezing cold and fog so thick we could barely see. Six of us cowboys gathered the cattle and counted them out the gate onto Highway 14. A highway patrolman passing by advised us to reconsider, warning us about the poor visibility. We told him we couldn't wait—the cattle were completely out of feed. We'd already been waiting several days for that weather to break, and it hadn't. We felt we just had to go. He said he'd stay with us as long as he could, flashing his lights, but if he got a radio call, we'd be on our own. Fair enough.

We could only see a few feet ahead at a time, but the cows lined right out and started walking east, single-file for the most part, and we cowboys did the same, horses almost skating on the icy shoulder of the pavement. We could hear quiet voices from the other riders; we could hear the cows chuffing and panting; and we could hear their hoofs striking the cold asphalt.

Beautiful, mystical, eerie, riding along in the white. Nerve-wracking, too. In addition to the patrolman, we had flagger pickups behind and ahead to warn traffic, but really they were useless—invisible in the fog—and we could only hope for the best.

We worried about the Dry Creek Bridge with its dark steel guardrails across a long span at the bottom of a hill. A bridge is a menacing structure for cattle; they don't want any part of a trap-like chute, close quarters, that scary echoing sound and then the rumble that starts when they get partway across. Even horses hate bridges and they tiptoe out, mistrustful of the footing. Our cowboys carefully pushed a few old lead cows out onto the bridge and jammed some others in behind them, tight enough that there was no way they could turn back, and we kept on pushing. The patrolman flashed his lights to stop the traffic out front, and Lester Werbelow, one of the men who owned the pasture, flagged behind us in his pickup. (Good of him, considering that the cows had eaten all the straw around his lighted nativity scene the night before.) What a relief when the last cows stepped out on the other end. The patrolman said he couldn't stay with us any longer, but the sun was peeking through by then and we were confident we'd make it.

A slow trip, and cold. Temperatures never got above freezing. We put the several hundred cows into a friend's pasture near Greybull just at dark and fed them a little hay. The next morning it began to snow and the pavement was slippery when we gathered them up and they started walking again. Their feet were sore and they were stiff from the cold, but these were strong cows. They'd left Shell Valley in May with their calves, they'd summered in the desert at Moneta, spent more than a month in Emblem, and now, January, they were ready to go home.

The cattle approached the railroad tracks near Greybull suspiciously, but fortunately there were no trains that day and the cows kept walking. Still ahead was a private bridge belonging to the M-I Bentonite Company—one long, narrow lane across the entire Big Horn River, used only for trucks and equipment. There's barely a little pipe side rail, no wings to funnel the cattle in, and a long, long drop down to the frozen, ice-covered river. Mart Hinckley, a supervisor at the bentonite plant and a born-and-bred cowman himself, had given us permission to use the private bridge, and he stopped the haul trucks for the time it would take us to cross. Meeting a truck at the wrong moment there could cause disaster.

We figured we'd have one good try at getting the cows out onto that bridge. We wondered if we should try to coax them with a little hay, or if it would be better to stay out of the way. Should we force them? Did we have enough help? What would we do if they just wouldn't go?

We needn't have worried. "That one old girl," our son Dan said later. "I was proud of her. She had her compass set on 'H' for Home. She never missed a step, didn't hesitate, just put her head down and kept on walking." The others followed. Beyond the bridge they strung out single file for a half-mile or so, like rough black beads on a long, long string. Frost on their backs, heads down, breathing hard, another five miles to go. But—headed home, home to Shell Creek.

When we reached the creek itself, ice ledges were frozen six inches thick. Men met us there with steel crowbars and even brought a tractor, prepared to open a safe crossing for the cattle and for our cowboys. The cows wouldn't wait; they plunged onto the ice, breaking it into huge slabs that broke and floated and bumped into the legs of our horses and the cattle, too. They stumbled clumsily into the frigid knee-high water, struggling as they climbed out onto the far bank. Somehow man and beast crossed safely, and we continued across a long flat to our fence line, just as the sky turned blue-pink against the sunset. "Sun dogs" hung like haloes in the frosty sky.

We closed the last gate behind the cattle at dark. They were visibly tired, but as the cowboys agreed, the "girls" knew they were home. Some lay down immediately, and others put their heads down to graze. We rode on to the barn, where we fed our weary horses and headed for the warm house ourselves. Tired, cold, relieved, triumphant. Another winter almost behind us, another spring ahead, and baby calves and green grass to come.

Home, sweet home.

18

Horses

When we brought the horses in from the range last year, my chestnut mare Brandy was not among them. She was old, of course, and I hope she simply laid down to die somewhere, finding the peaceful rest she deserved. I hope I never find her bones in some sad place, foot in a wire or a leg twisted under a rock. She was dear to me, and I cannot even guess the miles we covered together. I would hate to know she suffered.

Her distinctive type is seldom seen now. She and others like her, mostly Thoroughbreds, came into ranch work following the termination of the United States Army Remount Service, a predominantly Thoroughbred breeding program that provided reliable, useful horses for the cavalry in wartime. These horses were tall, long-legged, and comfortable to ride. They were tough and steady, capable of traveling long distances at a hard trot and when the program discontinued in the 1940s, ranchers quickly appreciated those qualities, seeking "Remount horses" for business purposes. Brandy came from those bloodlines.

Years ago, I rode Brandy for most of my ranch work. On one particular day, I rode her to look for a missing bull in the desert—a bull that had given us the slip when we gathered that big pasture the week before. My trail that day took me to a spot where Paul Daugherty, a well-respected old cowboy who worked for neighbors to the south, sat in his pickup truck beside the road. "I saw you when you started off the mesa," he said, leaning out his truck window, tilting back his hat in a gentlemanly fashion. "I knew it was you, knew it by the way that horse traveled, recognized that mile-eating trot. I could spot her a mile away. I always loved a horse like that. Just sat and watched her, waited for you here at the road. Stan's grandfather had horses like that. You don't see many of 'em, anymore."

Paul stepped out of the truck, leaned against the door, and wiped his forehead with the bandana from his hip pocket, then folded his arms across his chest. He propped a foot against the tire as I dismounted, both of us ready to share what news we could. Likely, neither of us would see another person that day. "I saw your bull—down on water. I just came from there. Hot as it is, he'll be close by," he said.

He offered me a drink of water from a jug in his truck, and we laughed together when he reminded me of the time he showed up just at suppertime at our family's camping spot on the mountain—exactly at the moment when Stan and the kids and I were dishing up a campfire-cooked supper. "I'd ridden by, early in the day," he said, "looking for some straggler cattle. I spotted the tent, and I knew it was your camp, knew it by that team of horses tied in the trees and the chuck wagon and that Dutch oven—big as a washtub. Seemed like a good place to be, come mealtime!"

"Not what everybody calls a vacation," I admitted. "It was a lot of work." We'd camped near Shell Lake so the kids could fish for a few days, close to a cabin where Paul spent most of the summer looking after the cattle from the south side of the Big Horn Basin. Instead of giving the family a jaunt to somewhere easy with a pickup truck, we'd used our draft team to pull the chuck wagon to that high country. Paul approved of Stan's view, that "the horses could use the experience." We hadn't had much experience at family vacations, either, and I remembered the kids' glee when lanky, smiling Paul came riding in from his camp a couple of miles away. When I last saw him, he looked the same as he did when he rode in to our campsite years ago: his Levi's and blue denim shirt are still softly faded to the same degree as the silvery gray of his hair and his hat.

Paul rubbed Brandy's ears as we stood, and she snatched bites of grass from the roadside as we easily fell into a conversation about horses. In my childhood, and in Paul's before mine, ranch horses served a purpose—to ride or pull or pack. Ranchers chose different horses for different tasks, some like my mare Brandy for big circles and others for sorting cattle or herd work, and they singled out gentle, dependable ones for "kids' horses." Ranching in early times did not accommodate fancy performance horses—"nothing but hobby-horses," Paul scoffed. Horseback sports and activities evolved during recent decades, making horses worth big money for rodeos and other competitions. "Bronc rides and horseracing were good enough for our fun—no such thing as team-roping or western pleasure, these games they have now," he said, twirling his old hat in his hands. "Some of these horses now," he added in his gravelly voice, "they're athletes, in top shape." He repeated an old cowboy adage: "Too much prosperity's bad for horses, just makes 'em fat and sassy. People get hurt that way. The horses don't get enough riding. They don't stay honest."

Most old-day ranch horses got plenty of riding, in contrast to the seldom-used and over-indulged ones that stand in stalls and pens today. Kids and horses made up a good combination, by learning and teaching. Ranch kids—our own, and my sisters and I when we were children—began with slow-poke, gentle "baby-sitter horses" which were friends and companions to us. As our riding skills improved, we advanced to livelier horses and we galloped from end to end of the ranch, laughing, playing, and racing, when we weren't doing cowboy work. Our horses got us where we wanted to go and let us have the freedom we craved. I treasure the memory of kids learning to ride on the likes of dependable old Bill or Peanuts or Zip. A 4-H kid's flawless performance at a fair or a rodeo is still irresistible to me, where kids are having fun, winning ribbons, pleasing the crowds, and enjoying their horses, too.

"Don't worry, he's as tame as a flea," our little Sara called out when her horse kicked up his heels. Nobody really knew how tame a flea might be, but we all laughed.

Dan, our youngest, described a Fox Trotter that belonged to a neighbor: "I think his clutch is slipping."

"My horse's name? I just call him Alpo."

"Frederic Remington? He's comfortable to ride, except when he bucks."

Sometimes the horses have the last laugh: "When he jumped sideways, I felt like he threw me past the moon."

Our kids early on knew their horses weren't "town horses," that while they were dependable for work or outside riding—"broke to death," the saying is—they might spook at the flag in a parade or protest if they heard firecrackers or the drums in a band.

Paul and I agreed: horses show their personalities just as people do. Some are kind and want to please; others are disagreeable, or lazy, or ill-tempered. Strong, healthy horses get the work done, so it's smart to keep them that way, feeding them well and treating them kindly. Sometimes, if you're lucky, a special horse reveals himself to be your true friend who earns and gives respect and appreciation by his loyalty and honesty.

Horses suffer wire cuts, broken legs, unsoundness, colic, illness, and injury, making for stories no one likes to tell. In one incident, our filly Banjo broke her leg—with all the family and crew watching. She stood in pain, confused, leg flopping, and we all turned away when someone said, "Get the gun." The beautiful mare Crazy Woman drowned in the Clark Fork River. Kind, gentle Fiddler was struck by a car on the highway. An expensive mare, Peso, turned out to be unsound, an evil-minded little thing, never the "dream horse" we wanted for our daughter Carol. A prize baby colt died when he fell through the ice on a pond. Our dear, dear old stallion Frosty suffered from old-age maladies until we ended his pain. Realties like these aren't happy endings. Horses can bring heartache, and often do. Paul shook his head, saying, "Seems like it's always the good ones that something happens to."

I told Paul about the afternoon I left a note on the table for Stan: "I'm at Steve's. I've got Trouble. Bring your rope—hurry." Stan grabbed his rope and roared his truck to the neighbor's, where we were branding calves. "What's wrong, what can I do?" he said, breathless, arriving in a cloud of dust. "Nothing. Did you bring your rope?" The horse I took to Steve's was one of ours, named Trouble, and I only wanted Stan to hurry because we were short of help. "Scared me to death," Stan said. "Where there's horses, there's trouble—not the first time I've heard that." A happy ending, that one.

There are other happy stories, too, plenty of those. Tim and I once put splints on the spindly legs of a baby colt we called Buckles because his ankles wouldn't hold him up. When summer came, we removed those hard plastic braces that reached from his hooves up above his knees, all the way to his little elbows, rubbed raw. We turned him out to the range then, with his mother, and watched him stumble away. Tim said, "Well, he's such a sweetheart, and so brave—the way he struggled in that miserable contraption. Probably he'll only get well enough to walk to the glue factory, but let's give him a good summer on mountain grass. Maybe he's got a chance." In the fall he trotted in, sound, his legs strong and straight! He's still traveling, twenty years later—classy and dependable, a loyal friend that provided many a good ride.

"You don't hear of a snubbing horse, anymore," Paul remarked.

"No, you don't," I agreed, laughing, "but I can tell you my snubbing horse story."

Stan was breaking a horse we called Widowmaker, one we'd bought. He turned out to be a complete knot-head. On the day we bought him, he pitched a fit when we went to load him in our trailer, and he got himself stuck in the poles of the chute. We had to take the corral apart to get him undone. We should have known right then that he wasn't going to amount to anything.

When the time came to break him, he didn't advance very well. "We'd gotten him to where you could get the saddle on, and that was about it," I told Paul. "Stan rode him in the round corral, but you can't stay in the corral forever." Paul nodded, knowingly. I continued my story, saying, "Stan suggested I should snub him for that first outside ride. I wasn't very sure of myself, but I thought I could do it."

"Snubbing" refers to a technique not used much today. The idea is to place a halter on the bronc while a helper on a broke horse leads the bronc, using a stout rope. The bronc rider on board remains passive until the colt gets used to the idea of a rider and relaxes. In theory, the presence of the other horse calms the bronc down. Old cowboy paintings often show such a scene, usually with a cowboy being flung off the bronc!

I knew that my good, reliable horse Scotty—a big, stout, black horse—wouldn't have any trouble bullying a colt around, so I chose him for my snubbing horse. Stan got mounted on Widowmaker all right, and we left the corral without any problem; I had the snub rope looped over my saddle horn, ready to dally if need be. We took off at a slow trot, with Stan advising me at every move, and I held Widowmaker on the right amount of rope to keep it from getting under my horse's tail and, at the same time, to prevent the colt from getting ahead of me. As the snubber, I wanted to keep the bronc out of my lap, too, which would turn a one-horse wreck into a two-horse wreck!

Widowmaker seemed nervous but manageable, and we jogged along okay for about a half-mile. He jumped a couple of times, but I got him under control without problem.

"When we started through some brush, though," I told Paul, "Stan's leg crashed against a tree limb, and Widowmaker spooked at the noise. He took a giant leap and came down bucking. I had relaxed, since everything was going so well, and I'd given the rope some slack. By the time I figured out what was happening, Widowmaker had a full head of steam and was way out ahead of me, and the rope slipped across my saddle horn. When I got my wits about me, I dallied hard and fast, and sure enough, Widowmaker came to the end of the rope. My horse, Scotty, sat back and hauled like he was supposed to, and he jerked Widowmaker so hard that the little bronc switched ends mid-leap—in a complete 180-degree turn. Stan's hat flew off, but he stayed on, although his direction had been completely reversed, too." Paul laughed, visualizing the scene. "Was he ever mad!" I told Paul. "And he swore like a trooper, but he didn't dare say too much for fear I might just turn him loose altogether. We finished out the ride. Eventually, we did get the horse broke, but he never was worth a darn, and we finally sold him. I did quite a bit of snubbing after that, and with a good horse like Scotty, I got pretty good at it."

Paul and I talked about what makes the *good* horse, the one that makes work a pleasure. A *good* horse is trustworthy and dependable, steady, and talented. My horse Pockets was one of those—a sweet, beautiful bay, a pal—he was "handy as the pockets on a shirt," someone said. On my happy days, Pockets carried me at a run, in fun and play, and I felt confident there was no horseback job the two of us couldn't do together. On sad days, I sometimes pressed my face into his neck and his soft skin, hiding tears I didn't want anyone to see. I counted on him to keep me safe and give me his all, and he always did. After I used him for many years, we retired him here on the ranch, to a life of rest and good feed. When it became apparent that he wouldn't have any more comfortable days, that he was hurting, we put him down, and buried him at the edge of a green pasture.

Fitz was a super horse, too, a fancy sorrel—well-bred, beautiful, smart, and agile. A cow could not get past him. When I had a chance to sell him for a lot of money, I didn't, and I wondered aloud if I'd made a mistake. Someone nearby said, "Nope. If you sell the horse, you spend the money—then you haven't got the horse and you haven't got the money. Keep a good horse like that one." In recent years I've been riding another good one, Mac, who makes his moves toward cattle beautifully, with dignity, and as precisely as a ballet dancer. He doesn't seem to love me, but he trusts me and I trust him, and he loves his work, and I love mine.

"It's as good as a paycheck," Paul said, "when that job gets done just right because you've got the right horse." Some days, a cowboy needs a traveling horse, and he rides that horse to cover some miles—a paycheck indeed when cowboy and horse trot strongly, purposefully toward a far-off ridge. On other days, it's a quiet, herd-working horse that is needed, or one who makes it safe and easy to rope a sick critter. Sometimes, a cowboy wants a sure-footed, stout horse to climb through rocks and timber. Another day, it might be a horse that won't spook when he needs to carry a baby calf in the saddle or step across a treacherous stream. "The world just looks better from the back of a horse. A horse is more than something to ride," Paul said. To back up his statements, he described one of his own favorites. "He doesn't look like much, but he never quit me, not once, in all these years and miles. He's just got *so much heart.*"

Paul and I shrugged, self-conscious in that quiet moment of sentimentality. He climbed back into his battered pickup truck and headed back toward his side of the range, waving his gray hat from out the window as he drove away. "Take care of yourself and that horse," he said, waving. "She's a good one."

———

A prayer, and thanks, to horses I remember:
Brandy Nibs Trixie Taffy Smiles Slivers Peso Banjo Murphy Minnie Joe Coquita Herb TheRoan Echo Bill Peanuts Apollo Fury

Flicka Tuck Dynamite Mike Pockets Hank Maizy Hollywood Savvy Twistywonder Blue Stampy Dan Reba Snip Calico Frosty Topper Whiskey BuckskinMare Sox Flash Fiddler BrownJug Smokey Belle Expo Touchdown Mac Moon Slammer BigOak Sugar Vanilla Red Wrangler Ace Handout Molly CindyJo CindySue Dolly Slim JohnChance Fitz Trouble PowderRiver Betty Nancy Alpo Lady JohnDeere Bearbait Tagalong Banana Fred Widowmaker Doc Scotty Pet Maude Jim Taffy Barney Babe Merle Pearl Zip SunlightSam Tiny Lucky Sandy CrazyWoman Sterling Irish Buckles Shorty Duck Blondie Ninja Chugwater OkeyPaint Stretch Slick Alpo FredericRemington Tari Pepper Cutter Dan LC Shooter Joe Dandy Flower Joker Sheila Kokomo Dude Stubby JJ Slick Jake Honey Charlie Ginger Stitches Billy Kicker Gypsy Bridget Pepper Alice.

19

◇—

Coffee with the Ladies

Springtime begins a rancher's year—calving season. Each day without fail for those several weeks, I saddle a dependable horse and ride to the meadow, where I check on the pregnant cows. The steady routine does not yield to circumstance of cold, snow, or wind—day in, day out, no matter the weather, I visit the cows in their huge bovine maternity ward. On this particular day, I hope to finish a little early. It's the day for the Shell Ladies' Coffee, and I'd like to go.

For a couple of hours each morning and evening I meander horseback through the cattle, watching for complications I can prevent or resolve. Some ranchers simply trust to survival of the fittest, but math always shows that a live calf is worth more than a dead one, and most ranchers conscientiously watch their cows during the weeks of calving.

When the weather is disagreeable, I hurry. On a bright, sunny morning like this one, I enjoy every moment. I've done this job for years, and my practiced observations of the details of pregnancy and motherhood can mean life or death for cows or calves. I'm familiar with the patterns of this large family; I know something about most of the cows although not everything about each one. Some are old friends—others are younger or "new girls."

As I ride past each cow I study details, imagining conversations. "Have you got a calf yet?" I say out loud. This cow's udder is full, but she switches her tail and wanders nervously in the trees. "Later today," I think to myself. "I'll check on her this evening."

"So *there* you are," I say, seeing a particular black cow; she has an odd split in her ear, and she's easy to recognize. I saw her last evening as she prowled in search of a cozy place to give birth. This morning in the willows, her baby lies nested in the tall grass; her bag hangs limp, showing he nursed, so I ride on. Nearby I'm relieved to see that a cow with enormous teats did calve earlier this morning, and her calf is big and strong and born with enough gusto to suck. I've been watching her each day, hoping for exactly that. "Ah, check that one off." It would have taken time I didn't want to spare this morning—to drive them to a chute at the corral and place the calf at her udder, to perhaps milk her teat to manageable size.

Today I hope for a routine check, a tranquil ride, no problems—not like yesterday when I found a cow who'd been trying to birth a backwards calf. She was down, exhausted; the calf was dead, and we had to pull it from inside her. Another day, a cow bumped her calf off a cut-bank as she enthusiastically licked the placenta off him; he couldn't get free until I hopped off my horse and climbed down to lift him. I did this carefully, though, because cows are instinctively protective, and the mother might have charged me, "taken" me, in rancher vocabulary. Sometimes a cow resists motherhood; she may refuse to claim her calf—or occasionally two cows calve near each other and they both want the same calf, leaving one abandoned. I can solve most of these problems, but it takes time—and if I find trouble today, I won't make it to the coffee.

I push my horse to walk a little faster, toward the fence corner where a lone cow stands beside her new baby. She watches suspiciously; as I near, I understand her wariness. An enormous golden eagle sits boldly in a tree nearby, fluffing his feathers in the warming sun, waiting to hop down for the afterbirth. When he sees me he stretches his wings wide, wide, but he doesn't fly away. The tiny calf's wobbly legs start working, and his mama coaxes him toward the safety of the feed ground.

Moving on, I find a steaming wet baby calf in a patch of brush, mother cow crooning to him, "Come to life, come to life, oh, you beautiful baby." Within minutes, she will instinctively lick him clean and nudge him onto his feet. She will pose *exactly so,* leaning to him, helping him find the teat for the colostrum he needs.

From across the meadow, a cow notices me and my dog; she decides to take her calf into hiding. She hums to him, "Come on now. *Here.* Stay close to me. I *mean* it. No fooling around." He's just learning to use his legs, but he obeys. She stops every few steps so he can lean against her, getting his balance.

I look away from the old, thin cow standing mid-field, waiting patiently for her calf to return. I stole him from her earlier in the week, but she hasn't given up hope; she waits where she last saw him. Over the years, she gave us many healthy calves and she took good care of them, keeping them at her side and leading them to green grass. This year, though, her milk was without nourishment and the calf grew weaker each day. Lest he starve, I took him to the corral to give him a new mother—a cow whose calf had died. I tell myself it was for the best, but her forlorn bawl haunts me. I'm not imagining her loss, her sorrow. It is real.

Last week I found a cow licking her born-dead calf, desperately trying to coax it to life. Confused and angry, she pawed the ground and shook her head at me. I didn't know what had gone wrong, but that day I had a happy ending: we'd found an extra calf at the barn, a twin, unusual for cattle. We could switch her dead calf with the live one if I could get her to the corral, half a mile away. I looped my lariat rope around the dead calf, dallied the rope to my saddle horn, and pulled the little body along behind my horse. The cow followed at a trot. "Is he alive? Oh, he's moving, he's alive!" she bellered. We put her in a chute where we used a veterinary tranquilizer to lull her while the hungry twin sucked, and by the end of the day, she took him as her own—a resurrection!

Ah—here comes Prancer. She walks with a peculiar high-step-ping gait; I recognize her from afar. She calved a few days ago back in the draw, and today is Parade Day, when she brings her baby out to show him off. "Lookee here," she seems to say. "Don't you ladies wish you had a fine bull calf like this one?" She marches regally, head high, not looking back, and he follows behind her.

Across Shell Creek I find another brand-new calf, still wet, and I recognize his mom, notorious for her bad disposition. She snorts protectively, shaking her head at me, so I leave them alone. She'll take good care of him. Uh-oh, here's a young cow, a little con-fused, kicking at her calf as he tries to suck. She's not eager for the burden of motherhood, I guess. I stay nearby to be sure she will accept him, and she finally does.

I ride over to an old cow I call Leona; her grumpy nature reminds me of a neighbor we once had. "At your age, you need a little more to eat," I say to her. Tomorrow I'll put her in another pasture where she'll have extra hay—a thin cow can't raise a fat baby. She hasn't calved yet, so she's okay for now.

I notice now that the cow I saw first has simply laid down and pushed her baby out in a few purposeful contractions; she's already on her feet and licking him off. I laugh, remembering our little boy Dan's observation the first time he saw a calf born: "Now how did he get *in* there, anyway!"

I've had a good look around, and I'll see the herd again this evening. I glance at my watch as I turn my horse toward home. It looks like I can make it to the coffee! I'm tired of mud and men and manure and overshoes, and I'm bored with conversations about hay and the cattle market. I hurry now, anxious for a shower, some pretty clothes, some girl talk. I want to laugh, to share my friends' lives, their joys and sorrows. There's nothing quite like coffee with the ladies.

20

Good Help

Life is easier with good help. That can be the highest praise—"good help"—a term that does not speak to gender, profession, or skills. It reveals itself personified in those individuals who pay attention, observe, and answer to a need—those who care and step forward to help.

A good helper often turns out to be a hero in a given moment: someone who hands me a bucket just when I need it, or someone who steps up in time to shove a shoulder against the wire gate, or the guy who makes the effort to help lift something heavy or pull a rope tight. Maybe it's the person who waits quietly, filling a critical gap near the loading chute, or the fellow who's content to "hold herd" instead of dashing about horseback, seeking something more high-profile than helping.

At the ranch, we've seldom hired cowboys—instead, we've used our money to pay the people who could fix fence, irrigate, or operate machinery. We liked to cowboy ourselves, and throughout the many years, we got most of the cattle work done with the help of our kids and our pals.

Among these pals were women who regularly, reliably provided excellent help, because they did what we asked them to do, whatever the job. They often brought a pan of brownies besides putting in a full day horseback; and most were good riders, eager to revisit their own ranch upbringing or simply enjoy a day outside. Some owned horses of their own, while others happily climbed onto whatever mount we could provide. Generally these women didn't have a big cowboy ego, and they needed no mollycoddling.

Certain friends—men or women—were our first choice to ask for help, but because they had responsibilities of their own, we'd often have to reach out further. We didn't need expert personnel, sometimes just a person to drive an extra pickup or stand in a gate or help gather a bunch of cattle in a pasture.

Our partner, Steve, and I agreed that some of the help Stan gathered up weren't really much help at all, and we teased him without mercy. Stan attracted "wanna-be cowboys," and when he got them to the job, he seldom looked back to see what became of them, while he rode off in a different direction. Steve and I would be stuck babysitting them when we already had a day's work to do with the livestock. Stan defended himself, claiming that Steve, too, brought some folks who barely had a heartbeat. "Cows can count," Steve would say, justifying his own selections. When the kids questioned his phrase, he explained that the cattle could tell when we had plenty of riders, even if some of them didn't actually provide much help.

Dawson was one of the guys who showed up for Stan's crew. He had retired from a successful professional career of some kind, and he had moved "Out West" for the outdoor adventures. He was ready to ride, he said, and he could hardly wait to use his horse, Roy. "RoyBoy," Steve called him, after listening to Dawson's endless commentary: "Good boy, Roy. Good boy, Roy."

On Dawson's first day with us, we arranged to meet at a certain spot. We noticed as we drove up that the door to his horse trailer was standing open, and the horse was inside, calmly switching his tail. Dawson was napping in the grass nearby; as we approached, he stood up and told us he couldn't get Roy unloaded. While we'd all had horses that wouldn't *load* without persuasion, I'd never seen one that wouldn't *un*-load. That should have given us a clue as to what to expect with Roy and his owner. Steve got out his lariat rope, smacked Roy on the butt and swore at him, and Roy hopped right out of the trailer, looking chastened. "Good boy, Roy." Roy definitely looked like his feelings were hurt, but Dawson got mounted, and away we went.

Steve and I spent the summer rolling our eyes at Dawson and RoyBoy. One day, we left the corral on a long circle, and Dawson was supposed to follow Stan, who was riding a bronc. Steve and I got our cattle gathered, then we rode to the ridge, expecting to see the two riders bringing cattle toward us. We saw one loose, saddled horse galloping hard toward us, and we both feared that Stan had gotten piled by that bronc, maybe even hurt. Instead, the loose horse was RoyBoy, with no Dawson in sight. We caught the horse and looked around; and then here came Dawson walking down the hill. He had gotten off to "use the bathroom," he said, and RoyBoy simply headed for the corral. Stan caught up with us and admitted he had not noticed when Dawson disappeared.

Nothing discouraged Dawson. He persisted in telling Steve and me what to do and how to do it. When I was backing a truck and trailer toward a gate to load some cows, Dawson stood behind the truck, making absolutely indiscernible hand signals. Deciding that I didn't know what I was doing, Dawson left his spot at the gate and walked toward me, saying he'd do the job—but he left the gate open and the cows, of course, bolted out. (Back to square one.) When Steve and I complained, Stan would say, "Aw, he doesn't hurt anything." One night at supper, though, after a trying day of herding cattle *and* Dawson, Stan got around to saying, "You know, that old Dawson is kind of starting to get on my nerves—how do you feel?" Steve and I looked at each other and exploded into laughs. "*That* guy. No help at all. Worse than a bad dog," Steve said.

Another of Stan's helpers, Ed, came along dressed in the highest cowboy fashion. He had his pair of fancy silver spurs and a rawhide rope; he had his own horse, too, but he never could get that horse to move out of a walk. He spent a lot of time talking about the really good cowhands he had known, and he implied that we weren't among them. He brought his new dog once, one he'd picked up at the animal shelter, convinced that it was a misplaced cow dog. The dog chased everything in sight and finally ran away, causing Ed to spend the rest of that morning trying to catch him. He never brought the dog again. Mainly, Ed just parked his horse at the edge of the work, sitting there looking picturesque—an improvement over his "help."

Dick Reed was a fellow from town who had built a new house near our BLM permit, filling his retirement dream of living in the country. Our neighborly relationship started out with a few rude phone calls between Dick and Stan. Our cows came pressing in from the hills during that hard winter, snuggling up to shelter against Dick's unfenced lawn and house. Stan told Dick that, by law, Wyoming is a "fence-*out*" state, and cows in the yard were Dick's problem, not ours. Dick was furious, but he was a good cowboy; he'd been a team roper and he had a well-broke horse. He saddled up and began to ride every day, taking the cows back out across the range several miles to the drift fence—a cold and miserable job, indeed. He and Stan didn't speak for about a month. One afternoon, Dick phoned, and Stan answered with some apprehension. Dick said wryly, "Stan, tomorrow's Christmas. Do you think I could have the day off?" They both laughed then, and we all got along fine after that. That winter was a tough one, and it wasn't long before the snow got so deep that we brought the cattle home to our own fields. Dick helped us.

Another one of the retired men was Bruce Kennedy, a local man who owned some nice horses. He was a successful businessman in his own right, but he was self-conscious about his riding skills, frequently saying with a smile, "Part of me's still a little boy in a Roy Rogers get-up, wanting to be a cowboy." He claimed, "It's like a club, a *cowboys'* club. You all just look different—your clothes, your hats, the way you ride. You look so comfortable. I'm just not a cowboy." He was definitely better-than-average help, and good company, and we didn't take time to declare whether or not he was a cowboy. We enjoyed it when he came along. He knew a lot about things we didn't, and we never tired of learning from him.

He ruefully admitted that he hated being bossed by women and, initially, he might not have realized that any family ranch operation probably has a woman or two who herds part-time cowboys as naturally as she herds cattle. Some of these women have surely "earned their spurs," as the saying goes, and they don't mind telling a rookie cowhand what to do, even if he does own seven newspapers or perhaps a bank.

Thus, it was hard on Bruce's ego the day we trailed our four hundred pairs to the high country. We were short of help, and in July's hot weather, with plenty of grass at hand, the cattle didn't want to move. Usually the experienced hands, regardless of age or gender, take the critical spots, and the other riders are spaced between, keeping the most help where the most trouble is likely to be. It was Bruce's lot that day to be near the drag, and things were getting busy back there: the calves didn't stay mothered, and they were starting to think about running back or going somewhere else—looking for chances to escape from the herd. Several of our veteran cowgirls had helped us with this drive before, and they saw what could happen. They worked their horses into a lather, relentlessly pushing the cattle—and Bruce—trying to keep the herd together. They let Bruce know what to do and when, and they didn't spare his feelings.

My son Dan and I were on the point that day, and we were surprised when Bruce showed up ahead of several riders along the big string of cattle. Dan asked Bruce if someone at the drag sent him, saying, "It looks like they're short of help back there." Bruce shook his head and spoke to Dan, kind of man-to-man, "Well, Dan, no, nobody sent me, but I just *had* to get away from all those *women* back there giving me orders. I just couldn't take it." Danny looked at him and said with a grin, "Well, you better not let Mom hear you say that, because if I go back to the drag to help *those* women, she's gonna be your new boss!"

After that, we teased Bruce unmercifully. The story he liked to tell, though, was of a day when Stan asked him if he could do a favor for us, since Bruce was going to be horseback near our cattle that week on a pleasure ride. Bruce said he figured we had finally recognized his skills and his stature and worth as a true cowboy. As he recalled, "I stood as proudly as a freshman ball player who was called off the bench for my big chance." Stan pushed his hat back, as Bruce told it, and sort-of scratched his head and furrowed his brow. Then he said, "Well, Bruce, I know it would be a little out of your way, but it sure would help me out. Do you think you could drop down there to the White Creek cow camp and feed the old tomcat?"

In truth, Bruce was part of the "cowboys' club" because he liked us and admired our profession, and he respected what it meant to us. Cowboying and horsemanship were only a small part of our friendship—and only a tiny part of the "good help" he was to us. He died suddenly, years back, but I think he'd have appreciated that as a character in a cowboy story that's told and retold, he made it into "the cowboys' club."

Membership to that club remains wide open, the only required credential being an interest in ranch work and people.

If such a club had a registry, it would include a long list of people who have helped us with one traditional drive, the "Bermuda triangle" of our cattle drives, the drive where *things go wrong*. It occurs when the cattle have roamed above the valley floor in what is known as the Mackey pasture—up where it's high and breezy, away from the bugs and dust of the lowlands. We gather the cattle from this huge sagebrush and grass range, then we drive them onto a long, long slope of several miles—uphill all the way through pastures that hold neighbors' cattle—before we reach our own private land again.

I paged through my journal recently, discovering events I'd forgotten, listed since 1973 as "the Mackey drive." Year after year, I wrote about those days without realizing I'd made a history of our friends and helpers, and as I read the old entries, I realized anew what those friends gave us in food, fun, and company, along with hard, hard days of work. That first year I wrote about the Mackey drive, Stan's dad was with us, and our children, youngsters at the time, and our nieces and nephews, and Mike Hinckley and her daughter and son. A sudden, terrifying thunder-and-lightning storm hit, scaring us all to death and pounding us with rain and hail. There was no cover or shelter on that plateau, and Stan told the kids to race for the cabin for protection—a mile or more beyond; he and Mike and I dismounted for a while, hoping we'd be safe and that we could hold the cattle together. When the lightning abated, we, too, left the cattle, hurrying to find the kids at the cabin. They'd shinnied out of their wet clothes and were warming themselves and giggling beside the fire they'd built in the fireplace.

I revisited entries from each year, some comical: "Dutch oven chicken for 'lunch' at five o'clock in the afternoon. That was a dumb idea." Another year, "Larkspur so bad at Hazel Early pasture that we couldn't move through there. You'd think we could have figured that out before now." I wrote that one of our friends "left a watermelon in Battle Creek, where we could eat it when we got there. We had to cut it with an ax, though." "Dan lost his beautiful black Stetson that he'd bought second-hand from Tim for $5. We couldn't find it." Once our nephew got lost, and we called off the work until we found him, delaying the drive until the next day.

One entry, in the late 1970s: "The Mackey drive. What an awful drive that is. I've about had it with cows and the whole bit. I don't like being pushed myself and I feel like it's been that way, all summer. I hate seeing the kids so tired and so much expected of them, and me, too."

In different year, I wrote, "A lot of riders and not much help." A huge run-back, several actually, played out all the horses, dogs, and cowboys. At the end of the day, a helper from town said, "Well, it gives someone like me an appreciation of what it's really like: it's not always fun."

In 1986: "Seventeen pairs of Levi's on the clothesline after we got back from the Mackey. This houseful of teenagers, changing clothes way too often! And my God, the groceries to pack lunches for all of that crew."

The following year, artist Harry Jackson (who was good help, having spent many cowboy days of his own) went with us, and I wrote, "Harry raised his arm and said: '*This* is the work of the Master. I could never paint this.'" And, he said, "Cowboys are great with kids. Except their own."

We relied hugely on our regular, dependable volunteers, and we always managed to have some fun. Sometimes, out-of-town people offered to pay us for the opportunity to come along, and we refused, politely. "No," we said. "Our work comes ahead of your comfort, and so we can't take pay." Many, many kinds of friends accompanied us on this drive, including several authors, artists, a librarian, a museum curator, a professor, a teacher, a medical student, a priest, visitors from South Africa and Australia, an electrician, a silversmith, an attorney, a geologist, photographers, and out-of-town cousins and relatives. In their own singular ways, most of them were "good help."

In 1996, when our older son, Tim, was in the middle of chemotherapy and our younger son, Dan, was getting the hay put up, Stan and I decided we could get that drive done by ourselves. We couldn't find help, but the cattle were out of feed in the Mackey, and we couldn't wait any longer. I wrote, "We stayed at the Mackey cabin the night before and gathered the cattle to the upper end of the pasture. Took a good dinner with us and a bottle of wine, so that part was fine. We took off early in the morning, but when the cattle hit the next pasture, and fresh feed, they just ignored their calves. We had one huge run-back. We each had one spare horse, which wasn't enough. The dogs got so tired, I felt sorrier for them than for myself. We barely managed to hold the herd together until the cattle were ready to travel. We did finally make it to our own gate, but I wouldn't want to try it again."

Through later years, staying overnight at the dilapidated old Mackey cabin became a traditional part of this cattle drive, shared by family, spouses, grandchildren, and friends. I discovered a creased envelope, folded and stuffed between journal pages, something I had written while on one of those rides, although I am not sure what year it was:

> The Mackey Drive—"As good as it gets."
>
> "I looked around when we were pushing up that long sagebrush slope, toward the last gate, and everybody was just where you needed someone. It was family this time, all good horses, all good cowboys. A mile of strung-out-just-right cows and calves. I realized it was one of the moments I've lived for."

The thirty years of journal entries about the Mackey drive share a common vocabulary, including a constant reference to "help." *Good help, not enough help, no help, extra help*—each of those phrases brings a face or an experience back to life for me.

No matter what the circumstance, I hope I've been good help to others, too, in friendship and in times of need.

21

Passing Through

"Maybe you heard," the man at the little Shell post office said. "I sold out. Sold the ranch."

"Yes, I heard that," I said. I had little to say, for in watching his dalliance with land and cattle, I expected it would end this way. It seemed clear from the start that he wouldn't last. He'd come to Shell to get away from some other life, and I always thought, "If you can't settle up in the old place, you can't settle down in a new one." We enjoyed his company, though, while he was here. I never tried to tell him what I expected—that life here would disappoint him, if what he sought was the myth and romance of riding the range and being one with the spectacle of cowboys, horses, and cattle.

Wyoming's history presents a pattern of people passing through, often moving on to a different place and some other life. Shell Valley has been no different. People came for their own particular reasons; some stayed and some went on, and while they were here, their ranks often provided stimulating, interesting additions to what might have otherwise been a staid and stodgy population.

In about 1979, a small panic occurred statewide—everybody suddenly wanted to live in the country, and small acreages were selling randomly. Citizens were concerned about the viability of these sales and the effect they'd have on schools, roads, and taxes. The State of Wyoming Legislature passed a land-use statute that feebly tried to create guidelines whereby counties could address rural sprawl, subdivisions, and access to small, isolated tracts.

The legislation provided for counties and communities to develop plans for local areas, an assignment that wasn't well received by many hard-core ranchers and farmers, but at least they kept their sense of humor. At a meeting at the Shell Community Hall, the land-use planning discussion got heated when it seemed that perhaps someone, somewhere, might control how private landowners dealt with their property issues: "Nobody's going to tell me what I can or can't do with my own land!" Of course, the word "zoning" terrified the crowd. One fellow stood up to say, "You old-timers just want to build a fence around the whole damned valley and not let anybody in or out!" From the back of the room somebody wise-cracked, "Oh, don't worry, we'd let *you* out!"

Many stayed, though, and brought new life to our community, even though they carved up nice farmland, built eyesore homes on windswept hills, and placed bright outdoor night-lights to glitter like ugly rhinestones all over the valley.

At any rate, the man at the post office had invested here, thinking maybe some scenic acreage in Wyoming would become good property and believing that he'd love the lifestyle of wide open spaces. He had a horse, and he bought some cows and a tractor and set about becoming a rancher. Never mind that his "ranch" was a fraction of a farm in Shell Valley, which had already been divided past any productivity.

We chatted awhile about the weather and topics of no consequence before he pulled the conversation back around, determined as he was to finish his say: "You know, I thought I wanted to be a rancher. I love horses and riding, and I love being out-of-doors. I love the West, really. I thought I wanted this life—tending to livestock and using the Lord's gifts of land and water.

"I believed ranchers to be gritty and capable, strong in character, admirable. To me, they were part of the land itself, heroic and hard-twist, as the saying goes. Fun-loving and rowdy, colorful. I admired that spirit and I reveled in hearing ranch histories. I envied all of that, and I wanted to be part of it. But that's not how it turned out."

I only nodded. At the old local coffee shop, the neighbors had speculated about how much money the guy had spent to buy the overpriced land, and on top of that, to build the new house. Now they'd be speculating about how much he had sold it for, wondering what money he made or lost.

I had met him and his wife when they first came to Shell Valley; they were friendly and enthusiastic, and full of ideas about the place they'd found. They would run exotic cattle, perhaps, or host bird-watch tours, or maybe hikes in the badlands. But after only a few years, they decided to move on.

"I'm not afraid of work," he continued, "but ranch life is *hard*. I found out that ranch people are hard, too, and bitter. They're tired all the time, and all they talk about is how busy they are, and they scoff at anyone who plays or hikes or takes vacations. Work is all that matters to them. They act like martyrs in a weird cult. I don't want to be like that." Odd, I thought, that he referred to my family and me as "they." And odd, too, his tone—challenging, defensive, combative.

"I just wasn't cut out for it," he said, righteously. "For one thing, I never made any money. Yes, I sold the ranch. Now I can sit on my porch and drink a beer or go fishing without feeling guilty because there's some job I'm neglecting."

I wished him well, and we parted. He moved away not long after, still wearing his cowboy hat. "He was just passing through," I told someone. As often happens, what was once a ranch became a web of roads and a collection of vacation homes for part-time residents who do not keep their fences fixed and who complain when cattle muss the lawns and patios.

Occasionally I remember the man's words, though, and I squirm a little at the way he described us: "hard" and "bitter."

My friends Marge and Ernie sold their ranch, too. They were the last generation of a long-time ranch family, and they moved into a nice home in a subdivision near Cody, where I stopped to have coffee with Marge one day. "How's it going, life in town?" I asked her.

"Oh, it's all right. I *wish* we could have stayed on the ranch," she said. "We couldn't, though. We just couldn't. We're old—hate to say it. I'm nearly eighty. We have no family to pass things down to. We spent a lot of time with young people in the community, but we didn't have any children of our own. I don't know how many youngsters passed through our place, working. We really enjoyed them. They often come back to visit and thank us for what we taught them. By ourselves, though, we just couldn't keep up with the work. And we couldn't hire good, dependable help; that's just not to be found, though we did try. So we sold, and here we are.

"Ernie was really tired, just tired of the load. And I was tired, too—of dragging garden hoses and fixing big meals and packing all those coolers and being the back-up man every time there was a crisis, and there was one every day it seemed.

"So we sold for a *very* good price. We ought to be happy. Maybe it's easier for Ernie. He goes down to the Irma Hotel every day to drink coffee with his old cowboy friends, and they talk about the cattle market and the new rancher people, and sometimes they play cards.

"For me, though, I don't have much to do in town. My life was Ernie and that ranch. I don't have many friends now and really, no purpose. Out there at least I could look out the window. I could watch the hawks that nested in the cottonwood tree every year and see the cattle grazing or the creek flowing; I could count the deer along the ditch bank. I loved that ranch, loved the demands of it, really. So did Ernie. I wish we could have stayed there. But we couldn't."

Perhaps a little embarrassed at her rush of words, she shook her head and quickly offered more coffee. I noticed that her hands were freckled from her years in the sun and that she moved quickly, efficiently, with purpose. Slender and attractive, she still seemed ready for any task.

"Who bought the place?" I asked. "Will they take good care of it?" She told me that the new owners had lots of money and enthusiasm. "They seem nice enough. They went to all kinds of range management symposiums and cattle-handling seminars, and they sure haven't asked for advice from *us*. Of course, they immediately built a brand-new fence around their entire place, whether the fence needed fixing or not, just like someone was going to steal the land itself. They built a monster entry gate to the ranch, and then a huge fancy house.

"We never spent much money on that kind of stuff," she sighed. "Seemed like the ranch always needed a new tractor, or the barn had to have a new roof, or we decided to put the money into the cattle market. I agreed with Ernie on those things. He would have built a new house, had I asked. I have a nice house now, a new car, and pretty clothes. But I'd trade in a minute to be young again and back at the ranch."

"It takes all kinds," I said. "A few people followed that cowboy fantasy about the time Willie Nelson and those celebrities playing country music were popular. You probably remember the wave of hippies and young folks—lots of them single, some of them just pot-smoking bums. Some of them were smart and fun, and some of them actually got jobs and they did stay."

"The investment people, with the big money, that's another story," Marge remarked. "Somebody's always saying 'They're not making any more land. Put your money in land. Land is always a good investment.'" I knew better, and so did Marge, considering

the glamour resort properties and subdivisions that had failed—
and even larger ranches that sold for prices that could never be
recovered in any agricultural use.

"The new people at our place? They're just passing through."
Marge chuckled. "I guess we all are, in one way or another.

"I hope I'm wrong," she continued, "but I think they'll tire of
ranching. It won't be what they expect. The scenery doesn't make
up for the profit they're probably not going to make. And those
folks are used to making a profit!

"Their children will go to fancy colleges and no one in their
family will want the steadiness, the endless work. These people
don't want to put the hay up *now*, when it's *ready*—they want to
wait until it's *convenient*—and the same with branding or moving
livestock. They don't know what they don't know, and they don't
want anybody telling them. I'm kind of glad I won't see what happens next."

"I know what you mean," I said. "It hardly seems worth trying
to get acquainted, the way they come and go. I told Stan that if our
new neighbors lasted five years, I'd drop by their place with a cake.
I could waste a lot of cakes, otherwise."

Marge laughed. "That's the truth. I never minded the work. I
got tired, I admit, and frustrated. But I loved seeing things turn
out right, and I felt such satisfaction in our achievements. We had
disappointments—hard years, good years—but over time, we
were happy in that life."

We chatted a little longer and finished our coffee. I promised
I'd come again, but Marge died later that year, and Ernie not long
after. I have driven past their old ranch since then; she was right—
the first buyers didn't last long, and the ranch sold again.

I wish I could tell Marge what happened next: real ranchers
bought the place. I don't know where they came from, or if they
are young or old. The cattle look good and the fences are tight; the
hay is in the stack and the old barn is still standing. Someone said

these new folks helped at the neighbor's branding, and I'm told the family has youngsters in the 4-H Club. The woman brought food to the community dinner, I hear.

Maybe they're not just passing through, I would tell Marge. Maybe they love and understand ranching, and maybe they love their work. Maybe they will be proud and tough, and maybe they will stay.

And maybe I'll drop by with a cake.

22

Ranching from the Grave

Our accountant tapped his pencil on the glossy wooden desktop in his office, and then he smiled and stood, pushing files and stacks of paper aside. "Congratulations. You got it done. A lot of ranchers don't. I've seen it too many times: it's easier to divide money than it is land, and it often comes down to that. I know you want the ranch to last forever, but there comes a day when that's out of your control. Let's celebrate—I'll buy lunch," he said.

An agricultural tax professional, Larry lives for his work, and the three of us—Stan, Larry, and I—have worked together as friends and business associates for more than twenty years. This day, we'd finished the legal paperwork enabling our ranch to pass forward to another generation of our family. Stan and I had focused on allowing the ranch to continue as a viable business—hard decisions to reach, since beyond business lies history, sentiment, and our love and loyalty to all of our children.

"Nobody lives forever," Larry continued, "and you can't run a ranch from the grave. Sometimes ranches just disappear because ranchers—like other people—don't want to face mortality. I hate it when all that's left is the brand on the gatepost or the coffee table, when a ranch is absorbed into a subdivision or some other land project—or another ranch—as if it never existed. I see that too often, in my work."

Larry's next words took a different twist: "Still, you might be surprised. More times than not, it isn't money that pulls a ranch apart. It's family dynamics." Like the priest or the veterinarian, an accountant knows how ranches and families work, and Larry went on to say, "There are ways to overcome financial difficulties, as you did. But if anyone thinks a big family can run a ranch together in harmony, he is probably mistaken."

Then he added, "You know, in some families, personalities just grate against each other like chalk on a blackboard—spouses, in-laws, children, siblings, grandchildren—and when the older ones decease, things fall apart."

"*Decease.*" Stan shook his head at Larry. "Die. You can say it."

"Whatever." Larry went on, locked onto his topic. "As you'd say, when the old folks *die,* a family reveals itself, and that can be ugly. One person might think he deserves a bigger share, or someone might just need money. Sometimes there's a bully, or a favorite, or a control freak. Maybe there's a drunk or a drug addict or someone with a personality disorder—ranches aren't immune to that stuff. And an older generation is understandably reluctant to put things into writing and make something legal, something that will stand. I'm sure you agree."

We donned our coats, walking toward the door. Stan began speaking slowly, "I guess you had to be there, like they say. I wonder if anyone will ever really understand what it was like, raising a family, paying the bills, working the hours we did to keep the ranch together. It was one crisis after another for years. We fought mud in spring snowstorms, ice and cold in the winters. Sometimes drought and only dust in the summers. Market busts, machinery that was old and worn out. Mary and I dedicated all our married life—more than fifty years—to this ranch, and I've been here my whole life. Legal documents don't express the way I feel about stepping aside, handing off the management and the responsibilities. I can't stop caring. But I'd hate to think it would just disappear, as you say."

At the restaurant, we talked in particular about the federal "death tax" that sent ranchers reeling when it became obvious that what a ranch was "worth," not its profitability, would determine the inheritance tax owed to the government. Lawyers, accountants, and legislators scrambled to find legal provisions that might better enable the succession of ranch property from one generation to the next.

"That was an awakening," Larry acknowledged, "but what I'm talking about is the difficulty in finding 'fair' and next, how hard it is to make that happen. It can be done. Too often, though, the older generation just refuses to take the steps to pass a ranch on. Maybe the patriarch tries to keep control for too long; maybe he ignores reality. Maybe the paperwork just seems too daunting. Or maybe he didn't get it done while he had the mental wherewithal; maybe he lost sight of his good intentions and time just ran out. Possibly he's simply angry at his own aging. Or he doesn't want to see hard feelings among the family. And, of course, there can be fatal accidents or illnesses that reset the clock for every generation. I've seen it all.

"Your papers are in place, though. In writing, signed and sealed. Be thankful—on some ranches, there are no heirs who *want* to continue. Count your blessings. Your ranch is in good hands. Believe me, I appreciate the difficulty of handing off a ranch—and a life's work. It's obviously a breathtaking, sobering experience."

Defying our good-natured protest, Larry paid for our lunch, and we parted.

Tired and relieved to have this day behind us, Stan and I headed home, both of us completely aware of the magnitude of our long-studied decisions and still wondering if we had done things right. We climbed into the car and turned toward the ranch, not in the mood for celebration or for grocery shopping or errands. "I'm glad we won't leave secrets or suspense for anyone—no dramas," I said. "It's not always that way."

Maybe to reassure ourselves that we had created an honest, peaceful succession of ranch ownership, we began to talk of other scenarios we'd seen. We thought of Don's ranch, where he handed work to his grown children, dangling promises in front of them, changing those promises frequently—never writing anything out, never formalizing any plan to pass that ranch on. "For those years, I felt like the donkey following a carrot on a stick, and finally we left," his daughter told me. Her siblings left, also. "We couldn't live there any longer without some security." Don could not manage without their help, and he struggled angrily; he leased the ranch to outsiders, and when he died, he left a vague will that only dealt with a few antiques. His second wife inherited his ranch. She eventually sold it, and the ranch "disappeared," as Larry would put it.

I heard another story at a school event when I unexpectedly saw a young woman whose family I'd known long ago. "That beautiful ranch sold after Grandpa died, to pay inheritance taxes. He didn't leave a will," she said. "It was a heartbreaker. He had never legally made any business arrangements. My husband and I wanted the ranch, but of course we didn't have the money. My kids won't grow up there, like I'd hoped. We'll never go back."

In another family we knew, the son assumed he'd inherit his parents' ranch as appropriate reward for his years of work. In a clumsy, last-minute revision of the will, however, the property went to a stranger who had befriended the patriarch at the end of his life. "It was a mess," Stan reminded me. "The family was disappointed and hurt at first, and then mad, and then they called in the lawyers, and when all of that was over, the ranch sold in useless parcels, a subdivision." I thought of the old British system, where the "remittance man" was paid to go away while the first-born inherited the kingdom. "Maybe that wasn't such a bad idea—at least everyone knew what to expect," I said to Stan.

"Leaving a family in a forced partnership, like an arranged marriage, doesn't seem right, either," I added. We talked about what our friend Tom told us: "I earned my share of that ranch," he said, "by working and living there with Dad for thirty years before he died. Dad's will had a provision that my sister could run some cows on the ranch, in her own name. She's a schoolteacher and lives two hundred miles away. Her family can't share the work and they don't want to. Kathy talks about what a wonderful childhood she had, growing up on the ranch, but she's not a child anymore, and she left the ranch when she was eighteen to go to college and make a career for herself. I stayed." I heard the frustration and bitterness in his words. "Dad meant well, I know, and he wanted to honor Kathy somehow. She's just not a part of this ranch, except sentimentally, and a ranch can't run livestock for free. I love my sister, and if I try to resolve this, I know it will cause hard feelings. I'll do it differently, when it's my turn."

Another rancher, Tony, told me his own story. We shared a bottle of wine on his porch, and he swallowed hard, a tear in his eye as he shook his head and smiled self-consciously. "My sister hasn't spoken to me since I sold the ranch. I did the best I could, but I was broke. Maybe she felt Dad entrusted me with the family legacy and I failed to hold it together."

Stan knew of other cases. "Look at the Andrews place: the ranch deed was recorded in *all* the kids' names when the dad died. They can't agree on *anything*. The place is empty—falling down— because they can't even agree to rent it out. The ones who need the money can't sell it, and the others won't. And further up the creek, that ranch was divided so many ways that no one could sustain a ranch; no one could make a living there."

"My ranch roots run deep," a woman friend explained to me. "I deserved a plot of land, I thought, and Dad went along with me in his will. It was a mistake, though. What I have is too big for a lawn, too small for a pasture. I can't enjoy it—it's too much work, and I have a job in town. It's all I can do to keep the fences fixed. I can't make any good use of it. I don't want to be a burden to my brother, who owns the bigger part of the old ranch, and I want his friendship more than anything. I want to share our family tradition. Maybe it's not possible."

"Frank had the right idea, maybe." Stan and I laughed, remembering our friends who owned a ranch on Horse Creek. "Our kids were grown, and they all had good jobs in town," Frank told us, grinning. "And the ranch was never big enough to support anybody." As time passed, Frank tired of the constant feuding with his millionaire neighbor. "Hell," he told us, "I couldn't stand the guy. We argued about ditch water and fence lines and even the cows on the road." One day Frank marched into the house, slammed the mail onto the table, and told Mrs. Mike, as he called her, "Well, you can quit worrying. I sold the ranch. I saw Neighbor Bigbucks today and he started carrying on and I told him he could just buy the sonofabitch if he had enough money. And he said he did, and I took the deal."

"Well," Mrs. Mike said slowly, elegantly lighting a cigarette, "I guess that would be a good price for your half. My half is not for sale." They did not sell after all, to Neighbor Bigbucks or anyone else, and the ranch is intact for now, still running a small herd of cattle.

Stan and I neared home, more than ready to leave this conversation.

"*I* used to think I wanted all of us *together* in ranching," I said. "I thought our ranch could get bigger and better that way, and we'd enjoy working with our family. That was *my* dream." I laughed. "You and me and four grown-up families? All of us? So independent and bossy and opinionated? It would be a nightmare, even if our ranch was the size of Texas. A nightmare. At least we got it sorted out, for one more generation at least."

Our son Tim is our partner, here with his family; he has dedicated his life to this ranch, as Stan and I did. Our daughters, Carol and Sara, live in Wyoming with their families, and they remain loyal to agriculture and ranching, and to us. At the time they were deciding on careers, there were few possibilities that enabled women to step into ranch management operations, and neither they nor we pursued that option. Our son Dan and his family realized before I did that our ranch was not large enough for all of us, and they now own a ranch in Oregon. Stan and I are proud that all our children are loving and respectful toward one another and toward the land. I believe it is important to them, as it is to Stan and me, for the land to remain "in the family," offering a touchstone and a sense of place to all.

"I never dreamed the time would go so fast, though," Stan added.

We turned into our driveway, thoughtfully watching the soft evening shadows bring down the day. A waddling pregnant cow crossed the road in front of us, so we slowed the car and admired the herd of cattle lounging nearby. Frisky calves bucked and played, too. "Feeling good, aren't they?" Stan remarked with a cowman's satisfaction.

Our road passes a small, lovely cemetery that sits at the edge of the ranch. Family members rest there, with headstones offering names and dates in lasting reminder of each generation. Some of those individuals ranched here and some came home to this place of rest. Mistakes are buried there, and hard feelings—achievements and satisfactions, too.

"Peace," I thought. "Rest in peace." But a ranch is a place of action—progress, births and deaths, growth and vitality. Chaos, sometimes—and excitement and purpose, change, and fun, too.

I glanced at the tall, white obelisks in the little graveyard, imagining that they sternly oversaw our work. I pointed in that direction, saying to Stan, "This is how it works." He looked at me, puzzled, and I explained, teasing, "Don't you see? That's how it's done: ranching from the grave."

23

Lucky Enough

When she was small, my granddaughter gave me a little wooden board that we hung on the wall of our cow-camp bunkhouse. "Here, Grandma, it's for you. I made it." In a slightly crooked tracing, she'd scratched the words to make a remembrance of the fun we had together: "If you're lucky enough to be in the mountains, you're lucky enough."

Summer was quickly turning to fall in the Big Horn Mountains. She'd been my riding companion and "right-hand man" during the previous weeks, but it was time for school to start, and she'd be leaving.

"I'll miss you," I told her. "You're a good little hand, savvy with horses and cattle, and a lot of fun—and you know your way around, now, too. You were a lot of help. I hope you'll be back next summer."

"Yeah, and I can saddle my own horse, now, too! I'll be back *before* summer, if I'm lucky," she grinned. "It was great being with everybody." Indeed, the ranch can be a jolly playground for our grandkids. It's a school of sorts, too. Of the nine grandchildren Stan and I have, only one of them lives here full-time. All of them, though, know how to ride horses and help with ranch chores; several of them learned to drive here at the ranch, and each of them can change a tire, put on jumper cables, or check the oil in the pickup. They all know how to pitch in when they need to, whether at a branding or in the house.

On this day, years after we hung the little board on the wall, I'm following a bunch of cows through a mountain park, and I'm feeling, as the little board says, "lucky enough" to be on the mountain and to be among the West's ranchers. Our fall calendar includes gathering livestock from high summer pastures and trailing them home to lower country. Our cattle have spent these recent months climbing to the tops of cool, breezy ridges, grazing on sunny slopes, and lying in grassy parks—summer vacation, you might say—as they prepare themselves and their calves for the winter ahead.

When we get the cattle home, it's like looking at our own school report card to see what shape they're in for weight or age, and to see how many are pregnant. Most important, though, is getting all the cows home. We try to count—how many go up and how many come down—but some will mix with neighbors' cows or end up where they don't belong. "You never say you've got 'em all," ranchers say. Instead, someone might ask, "How are you coming with your count? Are you getting close?"

"Count 'em in the gate, count 'em out. Count 'em every time," Stan says. A few years ago, he held tight to the idea that we were short seven cows. Seven. He knew it. He'd counted in and out. Seven, and among them a marker cow that had never shown up. The ranch hands and the family weren't so sure. We'd looked everywhere, and we were tired of hearing about the phantom seven, but Stan wouldn't give up. He hired a plane to look for them, and sure enough, he found them—across a fence, miles away from where they belonged, seven Diamond Tail cows just waiting for someone to open a gate so they could come home. I went with him horseback to retrieve them. We got them home, even if it was in January!

Finding all of the cattle in the fall is a big job for many ranchers. Some of them are where they're supposed to be, and others have wandered off to explore new ranges or mingle with other cattle. I like to imagine that they have made new friends, formed social clubs, found favorite places to eat, and traded babysitting. Now, as I ride toward a little bunch, they stand and stretch, bawling a bit to resist my intrusion. From a distance, I can see that I am short a calf; one of the cows is dry, I see when I get closer. I don't know her story, but her bag hasn't held milk for quite some time. Maybe the calf died, or maybe it has just gotten by as a bum somewhere. We'll figure that out at home.

Gathering the cattle holds some urgency because of the variables involved—where we'll find them, whether we'll get it done before hunting season, when the weather will change, who's available to help, and what the cattle market is like. Nowadays, most cowboy crews are family and friends—a pared-down version of the old roundups. Too often, we don't appreciate the chance to do this work surrounded by beautiful landscapes, riding with great companions. We hurry so we can get to the next job because there's always another one waiting.

I love this time of year, though, and the work that goes with it. The fall roundups bring ranchers together, neighbors who scattered—just like cattle—throughout hectic spring and hot summer, when feuds flared because of straying cattle, downed fences, or sparse irrigating water. Now, part-time and full-time cowboys meet, riding to locate animals that used the summer to wander. Under the warm autumn sun, we have time to share coffee from a thermos, and we can let the troubles and petty squabbles dissipate. It's a chance to exchange news, too, and useful information: "We have a cow and calf of yours; we saw her in the other pasture and we're gathering to the high corral on Monday. If you can't make it, we'll keep her safe for you," a fellow tells me.

Today, riding along with the sun on my back, I notice that my newly gathered little herd includes a pair that belongs to our neighbors. Their home ranch is fifty miles to the south, but our back fences border, and as the crow flies—or as a cow goes—fifty miles isn't far. In the ranching world, distance doesn't determine the bonds of friendship, anyway. Stan and I drove through dirt roads earlier this same week to retrieve a cow that had shown up at the neighbors' working corrals, and we finished out the day by driving down more dirt road to Hyattville so we could enjoy dinner with these friends at the local bar and grill. We'll need to call them and make a plan to return this pair to its rightful home.

It's an honor to have neighbored these folks for multiple generations. This ranch family has "hung on," since the late 1800s, when present-day matriarch Eleanor Walters Hamilton's grandfather arrived in that part of the Big Horn Basin. "Yes," she told me wryly, "we've stuck it out—our family—and now we have great-grandchildren on the place. Believe me, they all have as much advice for me as I do for them. I don't think anybody's listening. But somehow, we're still here." Nonetheless, Stan and I love a chance to be with them, feeling our own years racing past us, too. Another family, the Mercers, are still in place there on that south boundary—rangeland friends and neighbors for more than one hundred years.

I rode all day, every day, this week, horseback in Wyoming's glorious autumn, and my heart is full of appreciation for the opportunity I've had to live and work in this way. On a day like this one, I can ride a horse across canyon rims, against a sky so blue it hurts my eyes. "Looks fake," the cowboy said to me earlier today. Grinning, he added, "Purple mountain majesty, right?"

The small sign at the cabin says it all. Today I don't think of worries and doubts that wait at home. Instead, I reflect upon the enormity of our planet and my own insignificance in it, and I give thanks and praise to whichever God placed me here. I honor the beauty that surrounds me, smiling as I pass rockchucks whistling their hellos, and tipping my hat to view the hawks floating above me.

My horse tramps through crispy yellow leaves, piled like golden coins scattered by a careless king. Aspen trees blaze rosy pink, fiery orange, yellow; sagebrush branches blend in pewter gray. In deepest plum red, wild geraniums crawl amid the willow thickets, winding across cured grass. Trails braid down the hillsides in deepest, richest brown, and I imagine for a moment that they're veins in the giant wrist of Mother Earth. Below, creeks run clear, barely edged with ice earlier this morning.

Many of our old friends have passed away, and I miss them—people who rode with me in these familiar places—and I consider the changes I've seen. A few other family ranches continue on, but we're one of a few remaining islands of private ranchland now surrounded by resorts, subdivisions, and recreational businesses. Federal lands and corporate ownerships have their own sets of rules, and most of those rules don't acknowledge the communities or the landholders who were here first.

I yearn for the people—true ranchers—who could read brands, who made their living from land and livestock, who knew the geography of the country and appreciated the year of hard work that brought this fall reward. I miss my little-kid cowboys who grew up to take other jobs and live in far-off places. I wish for time to retell the old stories, to laugh and listen.

In this beauty, though, I can't stay blue for long. There are new friends to meet, new places to ride, and new stories to hear. It's supposed to snow tomorrow, and we'll be rummaging for ear-flap caps and overshoes and slickers, misplaced and forgotten since last year. The early storm's misery won't obscure the glory of that first snow—pure, white, wet, with patches of fog hanging like tufts of cotton from the canyon walls. The trails will be black mud, slick, and treacherous. The cattle, confused by this sudden transformation of their world, will finally peek out from snow-laden pine boughs, reluctant to leave shelter. My dog will nip and growl at them until they carefully tip-toe down the steep slopes.

I don't know how long, or if, or who among my family will stay here. I know times change, just as neighbors do. I have faith that someone will feel as I do, a love and respect for the heritage of the land itself.

Most of the cattle are accounted for now, and headed toward the lower pastures. I'll put my horse away for the day, and I'll wish I were starting over, maybe, young and bold. But for now, I pause for a moment, calm and at peace, satisfied with one day's work. How can I complain? There's work to do, and after all, I'm lucky enough.

◇— Acknowledgments

I extend appreciation and gratitude to my dear friends and family, who have provided the soul of this book. The list of people from our community who have shared so much is far too long to include here, but I am grateful to all of them.

At the least, I must thank my husband, Stan, and our grown children—Carol, Tim, Sara, and Dan, and their families—whose loyalty to ranching and ranchland continues. I thank my sisters, Betty and Nancy, who patiently provided historical and personal information as did the Sublette County Artists' Guild, by compiling many writings about the early days in the Green River Valley. The Green River Valley Museum has preserved the collection of Budd family photographs, and the American Heritage Center at the University of Wyoming has archived a reproduction of the original Daniel B. Budd diaries.

I thank Lynn Pitet, my friend and editor, for her time and her trained eye, and I thank those who enthusiastically provided photographs for my use: Melissa Hemken, Lee Raine, Sherre Wilson-Liljegren, Lori Hinman Dorr, and Bob Budd, among others. I remain grateful to teachers and mentors Wendell and Tanya Berry, and Teresa Jordan, who helped me believe I *really could do this*. And I thank Byron Price, Charles Rankin, and the very capable staff at the University of Oklahoma Press.